W9-ASR-327

SAFETY AND THE WORK FORCE

INCENTIVES AND DISINCENTIVES IN WORKERS' COMPENSATION

John D. Worrall, editor

ILR Press
New York State School of
Industrial and Labor Relations
Cornell University

Cover design by Joe Gilmore

Library of Congress number: 83-12706
ISBN: 0-87546-101-8

Library of Congress Cataloging in Publication Data
Main entry under title:

Safety and the work force.

Bibliography: p.
Includes index.
1. Workers' compensation — United States.
2. Industrial accidents — United States. 3. Occupational diseases — United States. I. Worrall, John D.
HD7103.65.U6S23 1983 368.4'1015'0973 83-12706
ISBN 0-87546-101-8

Copies may be ordered from
ILR Press
New York State School of
Industrial and Labor Relations
Cornell University
Ithaca, New York 14853

Printed in the United States of America
5432

CONTENTS

CONTRIBUTORS

John F. Burton, Jr., is a professor at the New York State School of Industrial and Labor Relations, Cornell University. He was chairman of the National Commission on State Workmen's Compensation Laws.

Richard J. Butler is assistant professor of economics at Brigham Young University. He is the author of works on the frequency and severity of compensation insurance claims.

James R. Chelius is director of the Bureau of Economic Research, Rutgers University. He served as staff economist for the National Commission on State Workmen's Compensation Laws.

Alan E. Dillingham is associate professor of economics at Illinois State University. He has studied the role of aging and sex differences in workplace injuries.

Stuart Dorsey is an economist with the U.S. Senate Finance Committee. He has taught at Western Illinois University and has worked for the U.S. Department of Labor.

William G. Johnson is professor of economics at the Maxwell School, Syracuse University, and professor of administrative medicine at Upstate Medical Center, State University of New York.

Robert S. Smith is associate professor at the New York State School of Industrial and Labor Relations, Cornell University. He has written extensively on economic aspects of workplace safety and health.

Michael Staten is assistant professor of economics at the University of Delaware. He is the author of a study of the effects of increased compensation on the workplace behavior of air traffic controllers.

John Umbeck is associate professor at Purdue University. He has written a number of works on the theory of property rights and contract choice.

John D. Worrall is vice president and director of economic and social research at the National Council on Compensation Insurance. Formerly, he was assistant director of the Bureau of Economic Research, Rutgers University.

TABLES

CONTRIBUTORS

John F. Burton, Jr., is professor at the New York State School of Industrial and Labor Relations, Cornell University. He was chairman of the National Commission on State Workmen's Compensation Laws.

Richard J. Butler is assistant professor of economics at Brigham Young University. He is the author of works on the frequency and severity of compensation insurance claims.

James R. Chelius is director of the Bureau of Economic Research, Rutgers University. He served as staff economist for the National Commission on State Workmen's Compensation Laws.

Alan E. Dillingham is associate professor of economics at Illinois State University. He has studied the role of aging and sex differences in workplace injuries.

Stuart Dorsey is an economist with the U.S. Senate Finance Committee. He has taught at Western Illinois University and has worked for the U.S. Department of Labor.

William G. Johnson is professor of economics at the Maxwell School, Syracuse University, and professor of administrative medicine at Upstate Medical Center, State University of New York.

Robert S. Smith is associate professor at the New York State School of Industrial and Labor Relations, Cornell University. He has written extensively on economic aspects of workplace safety and health.

Michael Staten is assistant professor of economics at the University of Delaware. He is the author of a study of the effects of increased compensation on the workplace behavior of air traffic controllers.

John Umbeck is associate professor at Purdue University. He has written a number of works on the theory of property rights and contract choice.

John D. Worrall is vice president and director of economic and social research at the National Council on Compensation Insurance. Formerly, he was assistant director of the Bureau of Economic Research, Rutgers University.

1 · COMPENSATION COSTS, INJURY RATES, AND THE LABOR MARKET

John D. Worrall

In the 1960s and 1970s injuries serious enough to result in workers losing time at their jobs rose rapidly. Concern with occupational safety led to the passage of the Occupational Safety and Health Act (OSHA) of 1970, which mandated a review by the National Commission on State Workmen's Compensation Laws of the system for compensating injured workers. During the 1970s the costs of compensating workers for injuries and diseases "arising out of and in the course" of employment nearly doubled, even after adjusting for inflation (see table 1.1). The increase in cost was the result of more liberal benefit programs adopted by the states, the increased frequency of claims, and, perhaps, the lengthening duration of disabilities suffered by injured workers.

Economists at many universities have been investigating the causes of the rising cost of workers' compensation and the increasing frequency of job injuries and diseases. Does the present system for compensating for occupational injuries and diseases promote an increase in injuries and costs? In 1978 more than two million families

I thank Steve Zrebiec for research assistance and David Appel, Monroe Berkowitz, Philip Borba, John F. Burton, Jr., Richard J. Butler, James Chelius, Stuart Dorsey, William G. Johnson, Janet P. Moran, and Robert S. Smith for helpful suggestions. I am grateful to two anonymous referees for helpful comments on the papers in this volume. The usual disclaimers apply. — J.D.W.

TABLE 1.1
Estimated Costs of
Workers' Compensation to Employers
and Benefit Index, 1972–80

	Costs (in $ millions)	Costs as a % of Payroll	Benefit Change Index
1972	$ 5,832	1.14%	100.0
1973	6,771	1.17	106.3
1974	7,881	1.24	112.6
1975	8,972	1.32	121.3
1976	11,045	1.48	129.3
1977	14,038	1.73	131.7
1978	17,000	1.85	136.7
1979	20,000	1.95	139.5
1980	22,000	1.94	143.3

Source: Price (1979, 1980, 1981); Price kindly made available the 1979 and 1980 statistics on costs and costs as percent of payroll in advance of their publication in *Social Security Bulletin* (1983). Figures for the benefit change index are from the National Council on Compensation Insurance.

received cash benefits from the workers' compensation program, and many of these families received cash benefits from other programs as well.[1] A group of economists have applied the tools of their discipline to attempt to determine if these benefits, and the manner in which they are paid, affect the safety and claim filing behavior of both employers and employees. This volume presents the results of their research, which were reported at the Seminar on Incentive and Disincentive Issues in Workers' Compensation Insurance, held at the Graduate Center of the City College of New York, on November 16, 1981, under the auspices of the National Council on Compensation Insurance.

The Workers' Compensation Programs

A workers' compensation program requires employers to provide cash benefits, medical care, and rehabilitation services to their em-

1. During the summer of 1978, 37 percent of the families receiving cash benefits were also receiving benefits from other transfer programs. Social security was the most important of these; 27 percent of the families receiving workers' compensation cash benefits were also receiving a social security disability or retirement benefit (see Worrall 1982).

ployees for injury or illness arising out of and in the course of employment.[2] The workers' compensation laws are state laws, which provide for mandatory coverage in forty-seven states, with New Jersey, South Carolina, and Texas being elective states. In those three states, employers who do not elect coverage forgo their right to the common law defenses against negligence suits and, as a consequence, most employers choose coverage. Federal employees are covered by the Federal Employees Compensation Act, administered by the U.S. Department of Labor.

The system emerged with a quid pro quo. In exchange for giving up their right to tort actions, employees are to get swift and certain payment from the workers' compensation program without having to demonstrate the employer was at fault. In exchange, employers enjoy limited liability for industrial injury or disease. Seamen and railroad workers are exempt from the state workers' compensation laws by virtue of the Jones Act and the Federal Employers Liability Act, and these workers retain their right to sue their employers.

The workers' compensation system is litigious in some states. This can erode the certainty of payment to injured workers that the system was designed to provide. Similarly, there have been many court challenges in recent years to the workers' compensation exclusive remedy doctrine, in which employees have brought actions against employers (Larson 1982).

Employers can fulfill their obligations to provide workers' compensation coverage by purchasing insurance from a private insurance carrier or from an insurance fund run by the state or by self-insuring. Eighteen states have state funds. Twelve of these compete with private insurance carriers for business and are usually referred to as competitive state funds. The other states have exclusive state funds, and private insurance carriers are not permitted to sell workers' compensation insurance in those states.

Firms that self-insure can usually do so by posting a bond or deposit of securities with the state industrial commission. Most firms are too small to meet the requirements of state law for self-insurance.

2. This brief explanation of the workers' compensation system is not meant to be complete. Those who would like a fuller treatment should see National Commission on State Workmen's Compensation Laws (1973). For the seminal treatise on workers' compensation law, see Larson (1978).

Group self-insurance is permitted in some states, and firms, usually in the same industry, can jointly self-insure.[3]

Claims made by injured or diseased employees fall into five categories:

1. *Noncomp medical,* or *medical only,* claims are claims that do not involve indemnity payments. Workers' compensation provides for virtually unlimited payment of medical benefits.

2. *Temporary total disability* claims are those claims for an injury or disease that prevents someone from working but from which full recovery is expected. Workers who have temporary total disabilities draw indemnity benefits after a waiting period, three to seven days in most states. If they are incapacitated for a period longer than that established by law, typically twenty-eight days, they receive retroactive benefits for the waiting period. More serious claims are also subject to waiting and retroactive periods. Temporary total claims are three times more common than the more serious indemnity claims, but they account for only 19 or 20 percent of the indemnity cost.[4]

3. *Permanent partial disability* claims are those indemnity claims for injuries that are expected, even after a period of healing, to result in a permanent impairment, functional limitation, or loss of earning capacity. Permanent partial claimants usually experience a period of temporary total disability. Permanent partial claims can be classified as major or minor claims, depending on the severity of the injury and indemnity claim, and as a group, these claims are by far the most costly in the workers' compensation system. They account for 65 percent of indemnity costs.

4. *Permanent total disability* claims are those resulting from injuries or diseases that prevent the worker from any work and are of permanent

3. The Chamber of Commerce of the United States publishes the *Analysis of Workers' Compensation Laws,* an annual that details, by state, the allowable insurance arrangements, the minimum and maximum benefits, the waiting period, the retroactive period, the replacement rates, and so forth.

4. The figures given are for different policy periods by state. They provide a good guide, however, to claim and indemnity distribution. The percentages of total indemnities for the four compensable disability categories for the last policy period reported by each state were, as of April 1982, temporary total, 19.4 percent; permanent partial, 64.5 percent; permanent total, 8.2 percent; and death, 7.9 percent. These calculations exclude Nevada, North Dakota, Ohio, Washington, and Virginia entirely, and the state funds for Montana and Utah.

duration.[5] These cases are relatively rare, two to three of every thousand indemnity claims, but they account for 8 percent of indemnity costs.

5. *Death claims* are those resulting from fatal injuries or diseases. Burial and survivor benefits are paid in these cases. Death claims are slightly more common than permanent total claims, and they also account for 8 percent of indemnity costs.

Indemnity Benefits

Indemnity (cash) benefits are paid in addition to any medical expenses for temporary total, permanent partial, permanent total, and death claims. As of January 1, 1982, most states paid 66.66 percent of wages for temporary or permanent total disability. Idaho and Washington, which paid 60 percent, New Jersey and West Virginia, which paid 70 percent, and Iowa and Michigan, which paid 80 percent of spendable earnings, were the exceptions.[6]

These replacement rates are nominal. The amount replaced is subject to minimum and maximum amounts in all states. The maximums have grown dramatically, partially as a result of recommendations by the commission. Twenty-six of the states have temporary total benefit maximums of 100 percent of the statewide average weekly wage; four states have maximums greater than 100 percent of the statewide average weekly wage; and the remainder have maximums between 66.66 and 100 percent of the statewide average or a specific dollar amount. The maximums are lower for permanent total than temporary total disability in two states, Utah and Wyoming. There are limits on the length of time a worker can collect temporary total disability benefits, but they affect very few cases. There are also limits on duration of benefits for permanent and death cases. These limits affect a relatively much higher percentage of claims. Indemnity benefits are not taxable, and as a consequence of this and the minimum and maximum provisions, the real rate of wage replacement

5. There are exceptions in some states, for example, the "working in pain" doctrine or the "odd lot" rule, which allows workers to work at temporary jobs or at occupations other than their normal ones.

6. Ohio paid 72 percent for the first twelve weeks, and 66.66 percent thereafter. See Chamber of Commerce of the United States (1981), chart 5.

can exceed 100 percent, or for high-wage workers, fall short of 66.66 percent.

Lump sum settlements (also called wash outs, redemptions, compromise and release settlements) are permitted in most states. They are more common in serious cases.

Death benefits are paid in all states. They are similar to those for total disability claims, with similar replacement rates, maximums and minimums. The percentage of wage replacement is reduced, however, in some states, for surviving spouses without children.

The Price of Insurance

The price that employers pay for their workers' compensation insurance depends, among other things, on their size, experience, classification, and insurance arrangement.[7] There are about six hundred workers' compensation class codes. These are used to classify businesses according to their primary activity.

Manual rates, basically the average of losses plus overhead, are based upon the experience of all firms in a given class and state. Manual rates are quoted in terms of $100 of payroll.

Firms that, on average, generate a premium of $2,500 at manual rate are subject to *mandatory experience rating.* There is some variation in this dollar amount and its date of adoption across states. In 1980, when the manual premium required for experience rating was $750, the average hourly earnings in the private sector were $6.66 (U.S., Department of Labor 1982), and workers' compensation as a percentage of payroll was 1.94 percent. A firm paying $6.66 an hour with three full-time employees in a classification rated at $1.94 per $100 of payroll would probably have qualified for experience rating in 1980.

Firms that qualify for experience rating have their manual rate modified based on their actual — as opposed to expected — loss experience. The experience of the most recent three years is used for the calculation of the modifier. The degree of credibility assigned to a firm's experience varies directly with the size of the firm. Large firms

7. The ratemaking process is a complicated one. An explanation can be found in *The Pricing of Workers' Compensation Insurance,* National Council on Compensation Insurance (1981).

generating $500,000 to $1,000,000 of premium are self-rated, and small firms generating $2,500 in premiums (and with lower statistical significance assigned to their experience) pay close to the group or manual rate.[8]

All firms that generate a premium of $5,000 or more are subject to mandatory *premium discounts.* The discount increases in steps with the size of the premium and is designed to capture economies of scale in overhead.

Firms that generate $5,000 of premium can also choose to purchase *a retrospective rating plan.* This option is usually chosen by larger firms. Retrospective plans are basically cost-plus insurance. The employer pays the loss costs, subject to bounds, and an insurance charge.

Finally, employers can receive dividends that substantially alter the net price of their workers' compensation insurance. In some instances, sliding scale dividend plans, based on the individual employer's losses, may be used. In others, flat rate dividend plans, the same dividend is paid to all policy holders.

Injury Trends

From 1972 to 1980 the incidence of occupational injury and illness cases involving lost workdays increased 20 and 30 percent for the private and manufacturing sectors, respectively. During the same period lost workdays per hundred full-time workers increased by 41 and 39 percent for those sectors, and the rate for nonfatal cases without lost work time fell dramatically. What was fueling the increasing frequency and lengthening duration of these more serious and expensive cases? Were some of them cases that would have previously not involved lost workdays? Would this explain some of the decrease in the number of cases without lost work time?

The increase in the incidence of lost work time cases during the 1972–80 period came on the heels of a large increase in the 1960s. The Occupational Safety and Health Act (OSHA) of 1970 mandated a new and expanded annual survey of work injuries and illnesses. Data collection under the new system began in 1971, but data for only a

8. *ABC's of Experience Rating* (National Council on Compensation Insurance 1981) provides a useful introduction to experience rating and calculation of the experience modifier.

half year are available for that year.[9] Old estimates, however, indicate that between 1960 and 1970 the incidence of injury cases increased by 27 percent in the manufacturing sector. Although the old and new incidence rate series are not comparable, two researchers have attempted to splice the series. Their results indicate that in the manufacturing sector injury frequency nearly doubled, increasing 86 percent, from 1960 to 1979 (Naples and Gordon 1981, table 1).

Proponents of OSHA believed that through its safety standards, inspections, and enforcement powers, the act would help to clean up the workplace, and reduce job-related illnesses and injuries. Some antagonists thought that OSHA would be a costly government intrusion into relatively efficient markets. The jury is still out on OSHA's effect on job injury.[10]

Workers' Compensation Costs

The act also created the National Commission on State Workmen's Compensation Laws. The commission was charged with the responsibility of investigating the state workers' compensation laws to determine if workers were receiving adequate, prompt, and equitable compensation for job injuries or diseases. It held nine hearings and submitted its final report (National Commission on State Workmen's Compensation Laws 1973), which made nineteen essential recommendations. Nine of these called for benefit changes. The U.S. Department of Labor (1981) has monitored the states' progress in complying with the essential recommendations and has found that from 1972 to 1980 states have made more progress in complying with the nine benefit recommendations than with the other recommendations.

Indeed, the rapid increase in benefits and hence employer costs during the 1970s was due in part to the commission's recommenda-

---

9. *Occupational Injuries and Illnesses by Industry, 1972* (U.S., Bureau of Labor Statistics, 1974) presents the first full year of data under the new collective system. BLS warns that the estimates for the six-month period in 1971 (published in BLS bulletin 1798) should not be compared with rates for later years.

10. See Mitchell (1982) for a review of research on the labor market effects of OSHA and other federal regulation. She reports on the data problems and mixed research results, but notes that "the best available firm-level evidence indicates that current practice has a small negative effect on workplace injuries" (p. 157).

tions. From 1973, the year after the commission's report, to 1980, the cost of workers' compensation insurance rose from $1.17 to $1.94 per $100.00 of payroll, a 66 percent increase in seven years. By contrast, in the twenty-year period from 1953 to 1972, workers' compensation costs rose from $0.94 to $1.14 per $100.00 of payroll, a 21 percent increase (Price 1981). Benefits grew from an index level of 100 in 1972 to 143.3 in 1980, a 43 percent increase.[11]

Today, benefits increase in virtually every state in every year. Nearly all states increase benefits automatically because their benefits are tied to their state's average weekly wage (Tinsley 1982). With increases in benefit levels, frequency and severity of injury, and increasing costs for workers' compensation came employer cries that their insurance rates were too high, regulators' demands to know why costs were rising so rapidly, and pressure on legislatures to do something.

Competition, Injuries, and Compensating Differentials

Economic theory holds that in a world of perfectly competitive markets, workers will be paid a compensating wage differential equal to the expected pecuniary and nonpecuniary value of risk borne by the marginal worker.[12] Furthermore, workers will be indifferent between buying their own insurance (the actuarially fair premium equal to the compensating differential) or the provision of workers' compensation insurance.[13] In such a world, changing the liability for injuries will not affect injury rates.

Benefits that an injured worker receives make an injury less costly to the worker. To the extent that employers pay for the benefits, accidents are costly to employers. In the world of perfect markets, the optimal number of accidents would occur. As employers sought to maximize profit and workers with their various tastes and preferences for risk bearing sought to maximize their utility, the accident rates and

11. See table 1.1.

12. I am simplifying here. Such markets would be in equilibrium when the marginal gains from all occupations that the worker could hold were equal. Injury risk would be only one factor contributing to the equilibrating process. For an excellent review, see Smith (1979a).

13. This implies zero transaction costs, perfect knowledge, and so forth. See Coase (1960) for the development of the Coase theorem.

compensating wage differentials in the various industries and occupations would be simultaneously determined. In such markets, attempts to reduce accidents below the optimal level would generate more cost than benefit. Similarly, forcing individual workers to bear less risk than they wanted to bear — and be rewarded for bearing — would reduce their satisfaction.

Obviously, the world is not a place of perfectly competitive markets, full information, and costless bargaining. Nonetheless, such a paradigm can be used to formulate and test hypotheses about the functioning of markets. The papers in this volume address a set of questions about markets that are less than perfectly competitive. In such markets, for example, workers and employers do not have full information about the costs and incidence of injury and disease. What happens if the acquisition of information is costly, and the information acquired is subject to uncertainty (Viscusi 1979b; Akerloff and Dickens 1982)? Will workers accept more (or less) risky jobs or lower (or higher) compensating differentials than they would if they had full information? In a world with less than perfect knowledge and with positive monitoring costs, will workers file fraudulent injury claims for benefits?

What happens when employers' costs do not reflect the full cost of injury because the insurance premium they pay for industrial injury and disease is set by recourse not only to their own experience but also to the experience of all firms producing the same general product? Will their incentives to reduce accidents and diseases be diminished? Will benefit increases in such circumstances lead to more risk bearing and more injuries and diseases, as well as more insurance claims?

Does the market currently pay compensating differentials for risk bearing? Are there the trade-offs between workers' compensation benefits and wages that economic theory predicts for a competitive market? These are the questions examined in this volume. As some of the evidence is mixed, I have attempted to reconcile the results.

The Effects of Workers' Compensation Programs

Benefits, Claim Rates, and Injury Incidence. There is a positive relationship between workers' compensation benefits and the rates for both indemnity claims and reported injuries. This finding by Butler (Chapter 3) and Chelius (Chapter 8) is a strong one. The same results have been obtained by Butler and Worrall (1983); Chelius

(1982); Worrall and Appel (1982); and Chelius (1977). But the research does not determine whether the increases in claims and reported injuries are the result of a reporting phenomenon or of more actual injuries resulting from additional risk bearing.

As workers' compensation benefits become more attractive, workers may have more incentive to file workers' compensation claims and their related injury reports. Or it may be that workers whose expected injury costs are reduced through higher workers' compensation benefits will be willing to bear additional risk on the job or will have less incentive to be careful. In either case, more injuries could result. It is quite conceivable that the positive association reported between workers' compensation benefits and indemnity claims is due to both the more frequent reporting of temporary total disability claims and additional risk bearing (or less careful behavior).

The estimates of the responsiveness of indemnity claims to benefit changes reported in some of these studies imply that up to half the increase in indemnity claims frequency can be accounted for by the benefit increases in workers' compensation insurance; John F. Burton, however, reports contrary evidence for New York State. But if the net effect of benefit change is positive and employee response is stronger, society may be buying adequate workers' compensation benefits with more injuries. Even if the actual injury rate is unchanged and higher benefits are simply inducing more indemnity claims, a portion of the rising cost of workers' compensation programs can be explained by this phenomenon.

Because benefits are set by the state legislatures, these bodies are understandably concerned with the increasing cost of their workers' compensation programs. The message of a growing body of research is that higher benefits will bring not only greater costs for the cases already being compensated but also more claimants and, perhaps, more work injuries and the costs that go with them.

Compensating Differentials. Workers are paid wage premiums, called compensating differentials, for bearing risk. This is found by Dorsey (Chapter 4) and Butler (Chapter 3).[14] Again, this has been found

14. Smith (1979a) reviews other studies, and Flanagan and Mitchell (1982) provide a review of wage determination and differentials. Ehrenberg and Smith (1982) devote a chapter of their most accessible labor economics text to compensating wage differentials.

previously in studies using different data sets, specifications, levels of aggregation, and time periods (Dorsey and Walzer 1983; Butler and Worrall 1983; Smith 1979).

Dorsey's study is the first to provide evidence that the premium for risk bearing is paid in both wages and fringe benefits. Previous studies have been unable to test for a risk premium on total compensation. His evidence indicates that the percentage risk premium paid is twice as large for nonwage compensation as for wage compensation.

The nonwage differential paid for risk bearing varies by type of fringe benefit and by the frequency and severity of injury. Increased risk of nonfatal injury, for example, is associated with increases in the probability of pensions, holidays and vacations, and insurance, and it is also associated with increased expenditure on those fringes. Increased risk of fatality, on the other hand, is associated with lower likelihood of pensions, holidays and vacations, and insurance and with lower expenditures for these fringes.

The finding that compensating differentials are paid for risk bearing strengthens the findings of other studies. We have evidence in the studies by Dorsey and Butler (as well as in Butler and Worrall 1982) that differentials are paid not only for elevated risk of death but also for increased risk of injury. Some earlier studies found such differentials only for fatal injury (Smith 1979; Dorsey herein; Butler).

Dorsey partitions the total compensating differential into its components, the wage differential and the fringe benefit differential. He finds that the percentage wage premium paid for a 1 percent increase in the risk of injury (i.e., the percentage compensating wage differential) was close to the percentage total premium (wage plus fringe) paid for a 1 percent increase in the risk of injury. As earlier investigators did not have data that enabled them to examine the total compensating differential, Dorsey's finding will provide some support for their use of the percentage wage differential. Dorsey also finds that the percentage fringe differential is greater than the percentage wage differential. Two things are apparent from these findings. First, the absolute wage differential (i.e., the dollar premium paid) will understate the total differential that a firm must pay workers for risk bearing. Second, the use of compensating wage differentials as proxies for the total compensating differentials may be less appropriate for classes of workers that favor higher fringe benefits at the expense of wages.

Evidence is also presented here that compensating wage differentials are paid for more than just the pecuniary aspects of risk bearing and that workers are risk averse. The implicit differential for an added day lost that Butler (Chapter 3) finds is high, relative to the daily wage for South Carolina workers for the period of his study. If this finding holds up in further studies, it will allow stronger conclusions about the incentive that employers have to reduce risk.

The Wage–Compensation Cost Trade-off. Employers make trade-offs between what they must pay for workers' compensation and the wages they pay their employees. Dorsey holds risk constant and finds a one-to-one trade-off. In other words, for every dollar increase in workers' compensation cost, there was a dollar decrease in wages paid. Butler lets risk vary over time and finds a less than one-to-one trade-off between workers' compensation *benefits* and wages. Dorsey's cost variable should be a good proxy for an aggregate workers' compensation benefit variable, because in order to pick up systematic variation in insurance expenses, he attempts to control for firm size with dummy variables.

This finding and those on compensating differentials indicate that labor markets are operating as they should. Although they do not exist in the paradigm labor market, workers are being rewarded for risk bearing and there are wage-benefit trade-offs. There are workers who are risk averse, have information about the job risk, and collect for bearing risk. Such a scenario sets the stage for questions about the role government plays in the regulation of workplace safety and health. Dorsey considers this in Chapter 4.[15]

Effect of Benefits on Labor Force Participation. The effect of scheduled permanent partial benefits on labor force participation is weak. Johnson (Chapter 7) finds a small, but statistically significant, negative effect of workers' compensation benefits on labor force participation in the year after a work accident and no significant effect three and four years later. He also finds that workers' compensation benefits have no significant effect on the number of hours worked by these injured workers four years after they were injured.

15. See Chelius (1979) for a more detailed discussion.

Johnson's findings are not inconsistent with the first three findings. They do tell us that, for the group he considers, there are virtually no *measurable* disincentive effects four of five years after injury. He is careful to point out that his sample is for scheduled impairments, one particular state, and one particular year (before the national commission report).

Most injuries result in claims of short duration, and less serious claims may allow greater flexibility in the adoption of a disability status. Most workers with less serious injuries would return to the labor market fairly rapidly. Johnson points out that a number of workers with permanent partial impairments leave the labor market and never return. These are people whose cases are the most serious and who have the least flexibility in the adoption of disability status. They are not as likely to be able to respond to market changes.

Response of Nonscheduled Benefit Cases to Economic Conditions. Burton (Chapter 2) finds in New York State that nonscheduled permanent partial cases are more sensitive to economic conditions than scheduled permanent partial cases. This is probably because in that state nonscheduled benefits are paid on the basis of actual wage losses, while scheduled benefits are paid on the basis of medical losses. Burton also finds a strong negative association of the benefit level and the frequency of nonscheduled permanent partial cases in New York. His measure of frequency, administrative closure in a given year, was the only consistent time series available.

The findings by Burton that the influence of benefit level on cases closed is negative is subject to several interpretations. The most appealing is that the employer incentives to reduce claims (or injuries) in New York State are greater than the employee incentives to increase such claims (or injuries). The nonscheduled cases considered by Burton are quite costly, and employers should have incentive to prevent them.

Burton examines the role that unemployment plays in the number of wage-loss cases in New York State. His findings are instructive: up to one-third of the increase in nonscheduled, permanent partial cases can be explained by the increase in the unemployment rate. His empirical results should be considered research in progress. He is attempting to extend the time period of his data series, add control variables to his model, and ultimately replicate Butler and Worrall (1983).

Many states are considering the adoption of wage-loss systems of workers' compensation insurance. Florida has adopted the system, and the costs of its system have fallen during its first few years of operation. State legislatures currently considering adoption of such a compensation scheme could avoid potential pitfalls by carefully considering the benefit structure of any scheme they plan to adopt. Burton's study offers a lesson cheaply learned: it is possible to have adequate compensation benefits and still minimize disincentive effects.

Filing Incentives, Monitoring Costs, and Benefit Levels. Filing incentives are related to the costs of monitoring, as well as to benefits. Staten and Umbeck (Chapter 5) find that air traffic controllers had incentives to alter job performance to support occupational disease claims (see also Staten and Umbeck 1982). As the cost of detecting fraudulent claims increases and the benefits of filing such claims and avoiding detection increases, there should be an increase in claims of certain types (e.g., stress, soft tissue). When a worker is presented the opportunity of having a higher level of real income by receiving a workers' compensation benefit than by working, given low risk of detection, some workers will opt for benefits. In an attempt to provide adequate benefits, the federal program may have set real wage replacement rates above 100 percent for some federal workers. Such benefits could turn the federal program into a thinly veiled early retirement program. Unfortunately, some federal workers would reach the optimum age for such a choice rather early in their work lives, and benefits they would receive would be increased with inflation over their normal working lives.

Most state programs do not adjust benefits for inflation, so the present values of awards in a state program are lower than those in a federal program. Because of these factors, the age at which workers covered by state programs are likely to file claims may be different than that of workers covered by the federal program. But the warning that Burton gives with respect to marginal replacement rates and the one that Staten and Umbeck give on average replacement rates apply equally to the state programs: expect increased filing in states that increase benefits substantially, especially if the probability of detecting fraudulent claims is lowered.

Experience Rating and Injury Prevention. Experience rating in workers' compensation has no measurable effect on employer safety. Chelius and Smith (Chapter 6) arrive at this surprising result using a rather crude measure of marginal premium cost. Note that they find not that the workers' compensation system has no effect on employers' incentives, but rather that the experience-rating program has no measurable effect. Their result, although preliminary, is an important finding and contrasts sharply with the findings of Butler, Chelius, and Dorsey. In the Butler and Dorsey studies, for example, there is evidence that employers should have incentive to reduce accidents: wages rise with injury risk, and employers pay compensating differentials that can be substantial. But the experience-rating plan produces no measurable effect on employer safety. It could be that the incentives offered by experience rating are not substantial, but recent research results by Victor (1982) indicate that the opposite is the case. Chelius and Smith are working with Victor on a follow-up study attempting to construct a more refined measure of marginal premium cost to retest their hypothesis.

Demographic Characteristics and Compensation Costs. The changing age and sex composition of the labor force, as well as the economy's shift of emphasis from manufacturing to services, have put downward pressure on workers' compensation costs. Alan Dillingham (Chapter 9) estimates that during the 1970–78 period, the change has depressed these costs by 5 percent — a reduction that has been far more than offset by increases in workers' compensation benefit levels and inflation.

Employer incentives to prevent injuries could change with changes in the industrial structure and the composition of the labor force. The average costs of injury and disease have changed in the past twenty years, and Dillingham's simulation indicates they will change in the next twenty years. He predicts that even if the recent benefit increases do not continue, demographic change and structural shift in the economy will force workers' compensation costs to rise in the 1990s.

Some Things We Do Not Know

None of the papers in this volume tell us unequivocally how the workers' compensation system affects the duration, or severity, of disability. Butler provides some evidence with a variable representing the number of days lost per employee due to injury. He finds a positive relationship between workdays lost and the workers' compensation benefit variable. His variable, however, includes a frequency component, and Butler's data were on closed cases. Depending on the differential rates of growth of durations of disability by claim type, his coefficients may be biased because he is not observing open cases.

Chelius reports evidence on severity. He regresses relative average number of days per case on relative benefits and finds a *negative* relationship. He warns that his measure also has a frequency effect that could confound his results. If benefit increases induce an increase in short duration cases, the average case duration could decrease, even if the duration of more serious cases were unaffected.[16]

Very little is known about the strength, if any, of employers' incentives. The studies here present a picture of a labor market that rewards risk bearing. In aggregate tests, the market passes muster. Ideally, future research should be concentrated on microtests of employee (and employer) response to workers' compensation incentives. As of now, there are very aggregated measures of risk and price variables (benefits).

Little empirical work has been done on the relationships among wages, workers' compensation benefits, and injury rates (or indemnity claims rates). This volume is a step in the systematic investigation of these relationships. I hope that the investigations here will stimulate further research on some of the questions that remain unanswered.

16. In fact, even if the benefit increases were unambiguously increasing severity — i.e., cases that would not have occurred (zero duration) occur (some duration) — and the duration of all other cases increased, severity measures based on average lost workdays per case might not pick up the increase in duration. Hazard rate studies are better suited to analysis of censured data (i.e., data on cases not closed). Studies of the duration of disability that eliminate censured data will have biased results if the ultimate duration of cases still open differs systematically from the duration of closed cases. Butler and I have used censured data for a study (1982) on temporary total low back cases, and we find strong evidence that workers' compensation benefits are associated with extending duration of disability for the group under study.

2 · COMPENSATION FOR PERMANENT PARTIAL DISABILITIES

John F. Burton, Jr.

Permanent partial disability benefits are the most controversial and troublesome aspect of the workers' compensation program in many United States jurisdictions. These benefits are paid to workers who, after maximum medical recovery, continue to have losses of actual earnings or of earning capacity as a result of their work-related injuries or diseases. Although permanent partial disabilities account for less than 25 percent of all cases paying cash benefits, they account for more than 60 percent of all dollars expended on cash benefits (Price 1979, table 5) and probably account for an even higher percentage of cases that involve serious controversies and litigation.

The discussions of terminology, existing compensation systems, and a proposed hybrid system are based in part on an earlier study by Burton (1978); a more extended treatment of some of the topics in these sections is included in Burton and Vroman (1979), which in turn draws on material from a research project supported by the National Science Foundation (NSF) under the title "An Evaluation of State Level Human Resource Delivery Programs: Disability Compensation Programs." A final report was submitted to the National Science Foundation in 1979. Monroe Berkowitz and John F. Burton, Jr., were principal investigators for the NSF project, and Wayne Vroman had primary responsibility for the wage-loss study component of the project.

I appreciate the comments on an earlier draft by Richard J. Butler, Arthur Larson, Eric Oxfeld, and John D. Worrall; the data collecting and computational assistance of Mark Pettitt and Dane Partridge; and the typing by Nancy Voorheis. I assume responsibility for remaining errors — J.F.B.

This paper investigates the incentive and disincentive effects of the various approaches to compensating permanent partial disabilities. It examines the criteria used to provide permanent partial benefits in several United States jurisdictions, with particular scrutiny given to the wage-loss approach of the Florida and New York workers' compensation programs.

A Conceptual Framework

An obstacle to a national analysis of permanent partial disability benefits is that the concepts on which benefits are based and even the terms used to describe the same concept vary among the states. This discussion assumes that a worker experiences a work-related injury or disease that results in permanent consequences. *Permanent* in this context means continuing beyond the point when maximum medical rehabilitation or maximum medical improvement (MMI) has been reached.

The first permanent consequence is an *impairment,* defined as "any anatomic or functional abnormality or loss after maximum medical rehabilitation has been achieved" (American Medical Association 1971). Examples of impairments are an amputated limb or an enervated muscle. The impairment probably causes the worker to experience *functional limitations*. Physical performance may be limited in such activities as walking, climbing, reaching, and hearing, and in addition, the worker's emotional and mental performance may be limited (Nagi 1975). *Disaggregated functional limitations* and *aggregated functional limitations* should be distinguished. The distinction concerns the level of affected activity. The disaggregated functional limitations are relatively specific consequences of the impairment, e.g., loss of degrees of motion of a toe joint. Aggregated functional limitations are more general consequences, e.g., loss of ability to walk.

Functional limitations, in turn, are likely to result in *disability,* which is used here to mean "inability or limitations in performing social roles and activities such as in relation to work, family, or to independent community living" (Nagi 1975). Two types of disability should be recognized: *work disability,* a loss of earning capacity or loss of actual earnings as a result of the functional limitations, and *nonwork disability,* loss of the other capacities implied in the broad definition of disability (such as limitations on various aspects of family life, including recreation and the performance of household tasks).

Work disability can be conceptualized as either *the loss of earning capacity* or the *actual loss of earnings*. In a strict sense, these two aspects of work disability must accompany one another. An actual loss of earnings only occurs if there is a loss of earning capacity. Nevertheless, the distinction is important because some types of workers' compensation benefits are based solely on a determination of a *presumed* loss of earning capacity, without regard to the existence or extent of actual lost earnings.

The extent of work disability and nonwork disability for a given worker depends not only on the extent of functional limitations but also on other influences. For example, the loss of actual earnings or decrease in earning capacity depends not just on functional limitations, but also on the worker's personal characteristics (e.g., age and education), the labor market conditions in which he or she must compete for employment, and the available sources of assistance, e.g., cash benefits and medical care.

The actual loss of earnings resulting from a work-related injury or disease is further analyzed with figure 2.1. The horizontal axis represents time, and the vertical axis measures earnings. In the case illustrated in figure 2.1, the worker had wages increasing through time from *A* to *B,* corresponding to the worker's higher productivity and general increases in the prices and wages. At point *B,* the worker experienced a work-related injury that permanently reduced his earnings. If the worker had not been injured, his earnings would have continued to grow along the line *BC*. Although these potential earnings cannot be observed, they can be estimated from information such as the worker's pre-injury earnings, age, occupation, and work experience. The worker's actual earnings can be significantly affected by the work-related injury. In this example, the actual earnings drop from *B* to *D* (which corresponds to zero earnings) and continue at this zero earnings level until point *E,* when the worker returned to work at wage level *F*. Thereafter, actual earnings grow along the line *F* to *G*. It is assumed that the worker's actual earnings never return to the potential earnings (line *BC*) that he would have earned if he had never been injured. Not all workers with permanent impairments or permanent functional limitations have this wage history. Some workers may return to their old jobs at the wages they would have earned if they had never been injured. Other workers may experience a total loss of earnings after their injuries. Thus the example shown can be considered an intermediate case for workers who experience permanent consequences of injuries or diseases, in that the worker has a partial but not total loss of earnings.

The measure of earnings loss used so far corresponds to what is labeled on the figure as *true* wage loss. This measure of wage loss is equal to the worker's potential earnings after the date of injury (*BC*) minus the worker's actual earnings after the date of injury (*BDEFG*). Although this definition of wage loss is appropriate for many purposes, including the assessment of the total consequences of a work-related injury, it is not the measure of wage loss typically encompassed in a workers' compensation statute. Rather, the statute will measure what is termed *restricted wage loss:* that is, the worker's earnings as of the date of injury, which were at level *B,* are projected into the future at that level, namely along the line *BH*. Then the wage loss that serves as a basis for workers' compensation benefits is measured as the difference between the line *BH* and the worker's actual earnings after the date of injury (*BDEFG*). As is obvious, the restricted wage loss is smaller than the true wage loss. Indeed, in the present example, there is a date where the actual earnings line *FG* crosses the line *BH,* which means there is no additional restricted wage loss after this date even though there is continuing true wage loss.

The final noteworthy point in the figure is the date when the worker has achieved maximum medical improvement.[1] Permanent disability cases are defined as those for which the worker has consequences that extend beyond this date, and the date of maximum medical improvement is when the worker's medical condition is considered stable enough that he or she can be rated for the purposes of determining the permanent disability benefits. In the case illustrated,

1. The date of maximum medical improvement (MMI) is the terminology used in Florida. The corresponding term in California is the date the injury is permanent and stationary. Virtually every state uses the concept, at least implicitly. It is the earliest date at which the worker can be evaluated to determine his or her eligibility for permanent disability benefits.

FIGURE 2.1
The Loss of Actual Earnings
for a Worker with
a Permanent Disability

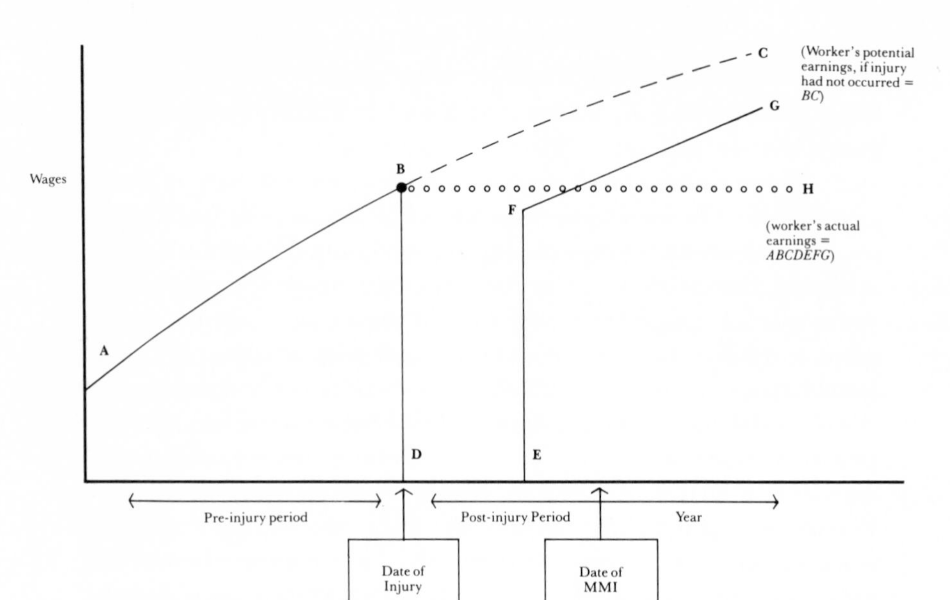

MMI = maximum medical improvement

"True" wage loss = potential earnings after injury (*BC*) minus actual earnings after injury (*BDEFG*)

Restricted wage loss = wage at industry projected into post-injury period without escalation (*BH*) minus actual earnings after injury (*BDEFG*)

the date of maximum medical improvement is after the date when the worker returned to work, a typical sequence.

Why Pay Permanent Partial Disability Benefits?

Why do we pay permanent partial disability benefits in the workers' compensation program? More precisely, which of the four consequences of a work-related injury provide the rationale for this kind of benefit? (For an extended discussion, see Burton and Vroman 1979.)

Commentators on workers' compensation generally have not considered the question using a conceptual framework as elaborate as the one used here, and so the answer to the question is complicated. A greater obstacle to ferreting out the purpose of the benefits is that, because the consequences of injuries are sequential and interdependent, a particular consequence may be endorsed as a basis for compensation because it serves as a convenient proxy for other consequences of primary concern. Thus a writer may argue that impairments should be compensated when the real concern is for work and nonwork disabilities that flow from the impairments. This indirect route to compensating disabilities may be chosen because impairments may be easier to measure than disabilities. Unfortunately, the commentators who favor payment for impairments do not always make clear whether this payment is for the existence of an impairment in itself or as a proxy for disability.

To the extent the rationale for permanent partial benefits is discernable, however, two schools of thought can be identified. One, exemplified by Arthur Larson (1973), considers work disability as the sole justification for workers' compensation benefits. He recognizes that some jurisdictions pay benefits on the basis of evaluations of the extent of impairment, but argues that when such evaluations are made, wage loss is conclusively presumed. The jurisdiction, in short, compensates the impairment because it serves as a proxy for work disability.

A second view of the rationale for permanent partial benefits is exemplified by the National Commission on State Workmen's Compensation Laws. The commission, in the 1972 report, concluded that the "primary basis" for permanent partial benefits should be work disability. The commission, however, also felt there was a secondary role for what it termed "impairment benefits," where the apparent purpose of the benefits was not only to compensate impairment per se but to use impairment as a convenient proxy for the functional limitations and nonwork disability that resulted from the impairment (Burton and Vroman 1979).

While not the only approaches to the question of compensability that have been offered, the views of the commission and of Larson are typical, however, in arguing that work disability should be the primary, if not the exclusive, basis for permanent partial benefits. Work disability as the dominant rationale for such benefits is used later in this study as an aid to evaluating the performance of the workers' compensation program.

How Do Workers' Compensation Programs Determine Compensation?

The previous discussion considered the theoretical rationale for permanent partial benefits. I now examine which of the consequences of work-related injuries and diseases are actually considered in determining the amount of workers' compensation benefits. Five approaches can be identified (for an extended discussion, see Burton and Vroman 1979).

Benefits Based on Impairments. The loss or abnormality aspect of impairment can serve as the sole or primary basis of permanent partial benefits. For example, the statutes in Washington and Oregon base some benefits solely on the extent of the impairment, and all workers with the same impairment receive the same dollar amount. The schedules for permanent partial benefits found in most statutes usually list injuries for which the extent of the underlying loss (impairment) is the primary but not sole basis for benefits. For example, the Michigan Workers' Compensation Act provides sixty-five weeks of benefits for the loss of a thumb, with each week's benefit set at two-thirds of the worker's average weekly wage (subject to a statutory maximum).

Benefits Based on Disaggregated Functional Limitations. The typical workers' compensation schedule contains, in addition to a list of injuries involving amputations, a list that relates benefits to the loss of use of specific body members. Thus, the New York law provides that "compensation for permanent partial . . . loss of use of a member may be for proportionate . . . loss of use of the member." The loss of use provisions are an example of benefits based on disaggregated functional limitations.

Benefits Based on Aggregated Functional Limitations. The California program has developed work-capacity guidelines to rate the severity of certain types of injuries considered hard to evaluate, such as spinal and heart injuries. These guidelines have eight levels, or plateaus, of severity, some of which are stated in terms of aggregated functional limitations. For example, plateau C is defined as contemplating that "the individual has lost approximately half of his pre-injury capacity for performing such activities as bending, stooping, lifting, pushing, pulling and climbing, or other activities involving comparable physical effort." This is an example of permanent partial benefits based on aggregated functional limitations.

Benefits Based on the Loss of Earning Capacity. In many jurisdictions nonscheduled permanent partial benefits are based on the loss of earning capacity. This loss of capacity results from the worker's functional limitations interacting with other influences such as age, education, and state of the labor market. Most statutes provide little or no guidance about how to evaluate these factors.

Benefits Based on the Actual Loss of Earnings. In a few jurisdictions, certain types of injuries are assessed in terms of their consequence on the actual earnings of the workers in order to determine the amount of benefits. This wage-loss approach is found in only a few jurisdictions, and even where found the execution is likely to be less than pristine. Several criteria can be used to identify a "pure" wage-loss approach. First, for an extended period of time after the date of maximum medical improvement, actual earnings are compared with potential earnings to determine if there is wage loss.[2] Second, if wage loss occurs during this period, then the amount of the permanent partial disabil-

2. The comparisons between actual and potential earnings could depend on monitoring by the state workers' compensation agency, or could depend on the worker or employer presenting data to establish the amount of wage loss.

ity benefits paid is related to the extent of the wage loss. Third, if no wage loss occurs during this period, then no permanent partial disability benefits will be paid.

A pure or virtually pure wage-loss approach is found in only a few jurisdictions. The 1979 Florida law that adopted a wage-loss approach for permanent partial disabilities and the New York nonscheduled permanent partial disability benefits are examined later. Pennsylvania also apparently uses the wage-loss approach.[3] There are no other verified sightings.

Michigan was cited by the report of the national commission (1972, p. 69) as using the wage-loss approach, and I have previously classified Michigan that way (Burton and Vroman 1979). Moreover, Michigan has recently been praised for its decision "to compensate for actual wage-loss only" (Devers 1981). But under the criteria just provided for identifying a pure wage-loss approach, Michigan appears to be a sullied wage-loss approach at best. The reason is the widespread use of compromise and release agreements (redemptions) that provide workers a lump sum settlement in exchange for the extinction of their right to claim continuing wage-loss benefits.[4] Use of a compromise and release agreement violates all three criteria for a pure wage-loss system and in effect converts the benefits into the approach based on loss of earning capacity, since the size of the lump sum settlement probably is based on an estimate of the likelihood and amount of future earnings losses.[5]

The five operational approaches to permanent partial benefits can be related to the distinction between scheduled and nonscheduled benefits found in most workers' compensation statutes. Each statute contains a list of injuries with a corresponding duration of benefits. The scheduled benefits are based on impairments or disaggregated functional limitations, the first two approaches described above (only

3. Arthur Larson, however, does not classify Pennsylvania as a pure wage-loss statute (1973, section 57.14(a), note 47).

4. During the field work for the NSF research project, a sample of files was examined for workers with permanent injuries who received Michigan workers' compensation benefits. Redemption settlements were found in a high proportion of all cases that involved nontrivial injuries.

5. For different reasons, Larson (1973) also does not classify Michigan as a wage-loss state. Indeed, he puts "it almost at the opposite pole from any actual wage-loss theory."

the unique California schedule uses other schemes in addition to the two approaches used elsewhere). Each statute also contains a general provision pertaining to injuries that cannot be rated by use of the schedule. One variant of nonscheduled benefits is based on the loss of earning capacity, the fourth approach described above. The other variant of nonscheduled benefits requires an evaluation of the seriousness of the permanent consequences of the injury compared to a "whole man." In practice, the whole man assessment considers impairments, disaggregated function limitations, and perhaps aggregated functional limitations in determining a rating. Either variant of nonscheduled benefits produces a rating (e.g., 25 percent), which is applied to a maximum duration corresponding to 100 percent (e.g., 500 weeks), to determine the duration for the nonscheduled permanent partial disability benefits (e.g., 125 weeks).

I will make three observations about the various approaches to compensating the permanent consequences of work-related injuries and diseases. First, although the primary rationale for permanent partial benefits is work disability and presumably the most compelling aspect of work disability is the actual loss of earnings, there is widespread reliance at the operational level on some of the other consequences of injury or disease as the basis for determining benefits. These operational bases such as impairment are presumably serving as a proxy for the actual losses of earnings that are expected to occur.

Second, with the exception of the wage-loss approach, the decision about the amount of permanent partial benefits to be paid is made after the medical condition has stabilized but before most or all of the actual wage loss for which the benefits are intended occurs. In short, permanent partial benefits are largely based on proxies for the expected actual wage loss that are assessed on an *ex ante* basis (that is, before the period when wage loss actually occurs). Whether the *ex ante* proxy approach works is considered in the next section.

Third, there is a tendency to confuse the various approaches, in particular the fourth and fifth approaches. A number of times I have heard a state's law described as involving a wage-loss approach, when the basis for benefits was not actual loss of earnings (the fifth approach) but loss of earning capacity (the fourth approach). The crucial distinction is whether the permanent partial benefits are paid on an *ex post* basis as actual wage loss occurs (the fifth approach) or are paid on an *ex ante* basis with the amount determined by an estimate of loss of earning capacity.

Does Workers' Compensation Provide Benefits to Those with Wage Loss?

Whether the workers' compensation program provides benefits to those workers with actual losses of earnings is one of the basic questions that was examined in a study sponsored by the National Science Foundation. (See Berkowitz, Burton, and Vroman 1979, part III; for summary of results, see Burton and Vroman 1979.) It made use of samples of workers injured in 1968 who received permanent partial disability benefits in California, Florida, or Wisconsin. Information was obtained from the state workers' compensation agencies concerning the workers' personal characteristics, such as age and sex, the nature and severity of the workers' injuries, the amount and timing of the workers' compensation benefits received by the workers, and the administrative procedures used to provide the benefits, including information on whether lawyers were used. The workers' gross earnings from 1966 to 1973 were obtained from Social Security Administration (SSA) records. Data on fringe benefits and taxes were unavailable.[6]

Information on each worker's earnings in 1966–67 was combined with earnings data for workers in a control group of the same age and sex to produce an estimate of *potential* post-injury earnings from 1968 to 1973 that the worker could have been expected to earn. The worker's *actual* post-injury earnings from 1968 to 1973 shown in the SSA records were then subtracted from the estimated potential post-injury earnings to produce an estimate of earnings losses caused

---

6. The proper measure of the worker's economic loss from a work-related injury or disease is the difference in net remuneration before and after the disability. "This comparison reflects factors that are affected by disability such as taxes, work-related expenses, some fringe benefits which lapse, and the worker's uncompensated expenses resulting from the work-related impairment" (National Commission 1972, p. 37). The absence of data on fringe benefits in the SSA earnings records means that losses of net remuneration are understated; the absence of tax data means losses are overstated. For a discussion of these and related issues, see Berkowitz, Burton, and Vroman (1979), chap. 17.

TABLE 2.1
Earnings Losses and Workers' Compensation Benefits, Male Workers Injured in 1968

	Potential Earnings, 1968–73	Earnings losses, 1968–73	Earnings Losses as % of Potential Earnings	Workers Compensation Cash Benefits, 1968–73	Benefits as % of Earnings Losses
Wisconsin					
Uncontested	$42,892	$ 2,519	5.9%	$2,150	85%
Contested	40,346	8,826	21.8	5,128	58
Florida					
Regular	37,023	190	0.5	1,372	724
Contested	32,733	6,898	21.1	3,534	51
California					
Informal	45,436	. . .	. . .	2,032	. . .
Formal	40,144	10,633	26.5	4,934	46
"Other WCAB"	38,683	7,166	18.5	2,905	41

Note: All figures are averages (means) for the workers in the sample. Additional information is contained in Burton (1978); Berkowitz, Burton, and Vroman (1979); and Burton and Vroman (1979).

by the work-related injury. (This corresponds to an estimate of true wage loss, as defined in figure 2.1.)

Table 2.1 shows the summary data for workers in the male samples for the three states. As an example, the Wisconsin male uncontested cases had mean earnings losses of $2,519 over the 1968–73 period, which was 5.9 percent of the potential earnings of $42,892 during the period. The cases that did not involve a hearing before an administrative law judge or the use of a compromise and release agreement (the Wisconsin uncontested cases, the Florida regular cases, and the California informal cases) experienced less wage loss on average than the generally more serious cases in the other samples. The most serious wage losses were experienced by workers in California who received formal disability ratings from the rating bureau: their mean losses of $10,633 represented 26.5 percent of the potential earnings of $40,144.

Workers' compensation benefits can be related to these earnings losses. Table 2.1 also shows the total of all types of workers' compensation cash benefits (after legal fees) received by the workers during the 1968–73 period and the percentage of earnings losses these benefits replaced. As an example, the Wisconsin male uncontested cases had mean benefits of $2,150, which were 85 percent of mean earnings losses. As the table indicates, the replacement rates varied considerably among the samples, ranging from 41 percent for the California other WCAB cases (cases that involved a hearing or a compromise and release agreement, but for which the rating bureau did not prepare a formal disability rating) to 724 percent for the Florida regular cases. The replacement rate for the California male informal cases can be considered even more extreme, since the sample had mean benefits of $2,032, even though the estimates show no mean earnings loss.

As divergent as these replacement rates are among the several male samples in the three states, the ability of the state programs to distribute benefits to those workers with actual wage losses is even less impressive when the samples are disaggregated to show the results for groups of workers who are similar in age and type and severity of injury and were provided benefits under similar administrative procedures. Tables produced for the National Science Foundation study indicate that even when groups of workers are compared, the Wisconsin, California, and Florida workers' compensation programs have serious equity problems. Workers with equal losses of earnings do not receive equal benefits — thus violating the test of horizontal equity — and workers with different losses do not receive benefits proportional to their losses — thus violating the test of vertical equity.[7]

The equity criterion is not the only test used in the National Science Foundation study to evaluate state workers' compensation programs. Another criterion is adequacy, which considers the average replacement rate for all workers in a state. A lenient test of adequacy requires that at least 50 percent of gross earnings losses be

---

7. The definition of equity used here implies that vertical equity requires a strict proportionality of benefits to losses. That is, a worker with twice the losses of another worker should receive exactly twice the benefits. More generally, vertical equity only requires that there be a consistent relationship between losses and benefits. Society may decide, for example, that benefits should increase less (or more) than proportionate to losses. The desired relationship between benefits and losses may be nonlinear, which is consistent with the use of minimum and maximum benefits. Evaluation of workers' compensation benefits becomes much more complicated when the test for vertical equity is something other than strict proportionality between losses and benefits. For an extended discussion of this point, see Burton and Vroman (1979), fns. 15 and 49.

replaced by benefits, and a more stringent test of adequacy requires that 66.66 percent be replaced. For the workers in the samples, the Wisconsin program replaced 75 percent of the earnings lost in the 1968–73 period, the Florida program 59 percent, and the California program 46 percent. For these three states as of 1968, workers' compensation benefits for permanent partial disability benefits were generally adequate.[8]

A third criterion used to evaluate the workers' compensation program is efficiency, which primarily considers the administrative costs of providing benefits under the workers' compensation delivery system, which includes employers, insurance carriers, workers, attorneys, and government agencies. Efficiency should not be defined as the cheapest possible delivery system, but as the least expensive system that can maintain a given level of adequacy and equity. Using that definition, the Wisconsin program appears to be preferable to both Florida and California. The benefits were more equitable than in the other two states, although still seriously inequitable; they were adequate; and litigation was much less prevalent in Wisconsin than in the other two states.

Possible Reforms of Benefit Programs

The approaches used as of 1968 in California, Florida, and Wisconsin to provide workers' compensation benefits for permanent partial disabilities appear to have serious problems. Although on average the benefits were adequate, the programs often failed to match the benefits to the workers with actual losses of earnings, thus causing serious inequities. Moreover, in Florida and California the delivery systems appeared to require an excessive amount of resources to achieve unimpressive adequacy and equity, thus demonstrating a lack of efficiency in those states.

8. The comparison of the benefits to the earnings losses assumes the sole purpose of the benefits is to compensate for work disability. To the extent the benefits are meant to compensate for some of the other consequences of injury, such as impairment or nonwork disability, the replacement rates will be lower. This will affect the adequacy evaluations made in the text and, probably to a lesser extent, the equity evaluations. This issue is examined in Berkowitz, Burton, and Vroman (1979).

The causes of these deficiencies are multiple and are analyzed in more detail in the report prepared for the National Science Foundation project (Berkowitz, Burton, and Vroman 1979). Among them are the reliance in most jurisdictions on a passive state workers' compensation agency whose primary function is to resolve disputes after they arise rather than to concentrate on procedures that prevent disputes. Wisconsin provides an outstanding example of a jurisdiction with an active workers' compensation agency that has been able to reduce litigation without sacrificing other virtues of the program.

Even in Wisconsin, however, the data suggest that equity is a serious problem. An important part of the cause for this problem appears to be the reliance on operational standards for permanent partial disability benefits that use proxies for actual wage loss to make decisions about the amount of appropriate benefits on an *ex ante* basis. Essentially, the permanent partial benefits are paid on the basis of an educated guess about the economic consequences of a particular injury — and the data from the National Science Foundation study indicate that the process is more guess than educated.[9]

What can be done to improve the performance of the workers' compensation program in dealing with permanent partial disabilities? Three possible approaches deserve mention: an improved *ex ante* system, a pure wage loss system, and a hybrid system.

An Improved Ex Ante System. An improved *ex ante* system would continue to evaluate permanent partial disabilities after the medical condition is stable and before the wage loss for which the benefits are intended occurs, but the criteria for rating the disability would be

---

9. The practice of using the impairment as a proxy for work disability has a long history of skeptical commentators. Addressing the Conference on Social Insurance in 1916, P. Tecumseh Sherman stated:

> Our American schedules are absolutely unprincipled. The amounts fixed for specific injuries have been determined not by estimating the resulting loss of earning power but by simple compromise between opposing political influences, thereby excluding any principle upon which to seek for finality or uniformity between the different States. Though the compensation fixed in such schedules for the minor injuries is sometimes fairly generous, that allowed for the major injuries generally is absurdly insufficient; whereas if either class of injuries is to be favored over the other, it certainly should be the major class. No allowance is made in such schedules for the variations in the loss resulting from the injured person's particular occupation, skill, age, etc. Such schedules also are all most insufficiently formulated (Sherman 1917).

improved. California provides a rough model for this improved *ex ante* system, because the rating system there does not use the anachronistic distinction between scheduled and unscheduled injuries found in most jurisdictions and there is a deliberate effort to incorporate some other influences into the calculation of benefits by the use of age and occupational adjustments. The main problems with the California schedule are that the standards for rating the medical condition are too loose and invite litigation; the ratings prepared by the rating bureau are virtually ignored by the parties and the administrative law judges when the cases become controverted; and the ratings produced by the system have not been validated against actual labor market experience. Although these deficiencies are serious, a state that wants to stay with a strictly *ex ante* approach to permanent partial benefits would nonetheless benefit from using California as a starting point.

The inherent faults of an *ex ante* system, however, can only be minimized, not overcome. One is the virtual impossibility of predicting as of the date of maximum medical improvement what will be the subsequent labor market experience of the disabled worker. Moreover, before the date when the size of the permanent disability award is determined, the worker has an incentive to exaggerate the apparent severity of the injury since that will increase the size of the award. The employer will try to downplay the severity of the injury, but the employer's incentive to reemploy the worker is muted because the actual labor market experience of the worker has only a limited influence on the size of the permanent disability award and consequently on the amount of the employer's costs.[10] Thus the incentives for rehabilitation and reemployment are restricted for both parties in an *ex ante* system.

There are some advantages to an *ex ante* system. The administrative burdens are reduced, since cases can be closed once a determination of the amount of permanent disability benefits is made. Moreover, once the case is closed, there is a strong incentive for the

10. The relationship between the size of the award and the amount of the employer's cost assumes the employer is self-insured or purchases insurance that is experience-rated. When an employer buys insurance that is not experience-rated, the firm has little incentive to find the worker a job if the earnings that result do not substantially affect the award the firm must pay.

employee to return to work since these earnings will not reduce the award. On net, I believe the disadvantages of the *ex ante* system dominate the advantages. But the record indicates that most states prefer this approach, and an improved *ex ante* system may be the best hope for reform in these jurisdictions.

A Pure Wage-Loss System. Another possible approach to reform is a "pure" wage-loss system, in which permanent disability benefits would be paid only if actual loss of earnings occurs. One variant was endorsed by the Interdepartmental Workers' Compensation Task Force in its 1977 report. Although benefits for nonwage losses based on the degree of impairment were supported as an optional addition, the basic permanent disability benefit was to be operationally based on the worker's actual loss of earnings. This *ex post* approach to permanent partial disability benefits is a possible solution to the problem of lack of equity since the purpose is to channel benefits to those who experience actual rather than presumed work disability.

But the pure wage-loss approach has its own problems. One is that, if the theoretical advantage of the *ex post* approach over the *ex ante* approach is to have any real meaning, cases will have to be kept open for long periods. The equity problem is not solved by denying benefits to a worker at the date of maximum medical improvement because he or she is experiencing no wage loss at the time and then denying benefits to the same worker experiencing wage loss three years later because the loss did not occur soon enough. A second problem, which has implications contrary to those of the first problem, is that determining the wage loss due to the injury is difficult, and the difficulty increases with time. On one hand, projecting the potential earnings for an individual worker (line *BC* in figure 2.1) is a formidable task. On the other hand, any shortfall of actual earnings below potential earnings in the period after the date of maximum medical improvement may be due not to the injury but to the myriad other factors that can adversely affect earnings, such as a plant shutdown that affects injured and uninjured workers alike. A third problem with a pure wage-loss approach is that if any amount of wage loss is compensated, no matter how small, the administrative burdens can be substantial.

A fourth problem with a pure wage-loss approach is that since the amount of benefits after the date of maximum medical improvement is dependent on the extent of wage loss, there is less incentive for an employee to return to work (since that will reduce benefits) than in the *ex ante* approach in which the amount of permanent disability benefits does not decline if the worker has earnings. The possible work disincentive effect is aggravated if the benefits are tax-free and if a very high proportion of wage loss is replaced by benefits. A fifth and related problem is that many workers are still in a rehabilitation phase at the date of maximum medical improvement, and the disincentive effect of wage-loss benefits can be particularly acute for these workers (see Keefe 1981). A sixth problem is that if the only type of benefit payable after the date of maximum medical improvement is a wage-loss benefit, then some workers — those with no wage loss after the date of maximum medical improvement — will receive no permanent disability benefits, even if they have such serious permanent impairments as the amputation or total loss of use of an arm.

Although these problems suggest that a pure wage-loss approach is likely to find few adherents, there are some advantages to the approach. The potential for achieving equity is much greater in a wage-loss system than in an *ex ante* system. Moreover, while the incentive to return to employment after the date of maximum medical improvement is less for the worker in a wage-loss system than for the worker in an *ex ante* system, the employer's incentive to reemploy the worker is greater, since the payment of wages translates into lower benefit payments. On net, I believe that, compared to the *ex ante* approach, the disadvantages of the pure wage-loss system dominate the advantages, but the case is not compelling.[11]

A Hybrid System. A third approach to reform of permanent partial disability benefits would be a hybrid system, combining some of the best elements of the *ex ante* system with some of the virtues of the pure wage-loss approach. Operationally, such a hybrid system could begin

11. A wage-loss system can be àdequate, inadequate, or excessively generous, depending on the design of the particular system. It is not evident whether the wage-loss approach or the *ex ante* approach involves more administrative costs. Under the wage-loss approach, hearings to determine the amounts of permanent disability benefits can be eliminated when the workers experience no actual losses of earnings. The wage-loss approach, however, requires cases to be kept open for extended periods, and benefits must be periodically recalculated if actual earnings fluctuate.

with a comprehensive set of standards for rating impairments and functional limitations. These standards should be issued by the state workers' compensation agency to replace any schedules included in the statute. The rating standards would emphasize objective factors (such as limited motion) rather than subjective factors (such as pain) in order to reduce the opportunities for controversion.[12] Use of these standards by a disability evaluation unit within the state agency should be obligatory before permanent partial disability benefits are paid or the case goes to a hearing, and the unit's rating must be accepted by the administrative law judges, agency appeals boards, and the courts unless the rating is clearly defective. The rating produced by the evaluation unit can be considered the standard rating.

The standard rating can be used as a component of several benefit systems that are described in the final report to the National Science Foundation (Berkowitz, Burton, and Vroman 1979, chap. 18). One system would contain two types of permanent disability benefits: presumed disability benefits (type I) and actual disability benefits (type II). Both would be paid in addition to any temporary disability benefits. The type I benefits could be based on the standard rating, or the standard rating modified by several objective factors that usually influence the extent of work disability, such as the worker's age and education, and that provide approximations of the influence of personal characteristics on a worker's post-injury labor market experience, given the seriousness of the injury. The factors could be incorporated by use of an objective formula, such as the one used in California, and their influence on the standard rating would not vary from case to case because of subjective assessments by an administrative law judge about their likely impact for particular workers.[13]

Once the modified rating (or standard rating, if the jurisdiction decides not to incorporate other factors into a formula) is determined, the amount and duration of the presumed disability (type I) benefits are calculated. The preferred approach would pay the type I benefits

---

12. The American Medical Association's *Guides to the Evaluation of Permanent Impairment* (1971) are a possible starting point for the development of state standards.

13. The formula incorporating the adjustment factors would be designed so that on average the standard ratings would translate into the same level of modified ratings. In other words, modification factors would be as likely to increase as to decrease the standard ratings.

for a fixed period for all cases, for example, six months, with the amount of the weekly benefit varying as a function on the size of the modified (or standard) rating.[14] Thus a 5 percent rating would produce six months of benefits, with the weekly benefit equal to 5 percent of the worker's pre-injury wage, subject to a maximum benefit of 100 percent of the state's average weekly wage. It may be desirable to have the fixed duration for the presumed disability benefits be longer than six months. Important principles are that these presumed disability benefits are paid without requiring the worker to demonstrate any actual loss of earnings and that the period of presumed disability benefits is long enough for most workers to recover their strength and earning capacity so that they are unlikely to be experiencing actual wage loss by the end of the period.

Under the hybrid system, a worker could also be eligible for type II benefits if actual wage loss continued after the type I benefits had expired. The worker's potential earnings (defined as the worker's pre-injury earnings and preferably increased through time to reflect changes in wages or prices) could be reduced by 20 percent, for example, to produce the worker's threshold earnings. If actual earnings are less than the threshold earnings, the difference is the earnings shortfall. Then the type II benefits could be calculated as 80 percent, for example, of the earnings shortfall.

In order to gain experience with the plan, particularly in terms of the potential costs, several limitations on eligibility for benefits seem appropriate, at least during the initial years. For example, eligibility for type II benefits must be initially established within two years after type I benefits expire. Also, the burden of proof to demonstrate that the earnings shortfall was due to the work-related injury or disease would be on the worker if the adjusted disability rating were between 11 and 40 percent, and on the employer to show the contrary if the rating were over 40 percent. The worker would not be eligible for type II benefits if the modified ratings were 10 percent or lower.

These rules and percentages for the type I and type II benefits are meant to be illustrative and would have to be modified for each

14. An alternative approach would make the duration of the benefits dependent on the size of the modified rating (for example, each percent of rating would result in four weeks of benefits), and the weekly benefit could be 66.66 percent of the worker's pre-injury wage, subject to the maximum benefit of 100 percent of the state's average weekly wage.

jurisdiction in light of prevailing economic, administrative, and legal standards.[15] There are, however, seven general guidelines for a hybrid system that seemed relevant for almost every jurisdiction.

1. There should be a lag between the date of maximum medical improvement and the initial date of eligibility for wage-loss benefits. The lag should be at least three months and preferably six months or more. This guideline is designed to deal with the disincentive problems for workers still in the rehabilitation process.

2. At the date of maximum medical improvement, all workers with a permanent impairment of any type should receive a scheduled award. The impairment should be evaluated by an objective standard, such as the *Guides to the Evaluation of Permanent Impairment* (American Medical Association 1971). The standard impairment rating may be modified by objective factors, such as age and education, but this should be done by a formula and not by a subjective evaluation of the facts in the individual case. The standard or modified rating should be used to determine benefits that are paid until the initial date of eligibility for wage-loss benefits. These presumed disability benefits can be viewed as incorporating payments for impairment and nonwork disability, as well as work disability. The payment of these scheduled benefits for all workers with a permanent impairment should rectify the inequity of some workers receiving no permanent disability benefits under a pure wage-loss approach.

3. Wage-loss benefits would be paid only to workers with at least a modest amount of lost earnings. A required amount of wage loss should be at least 15 percent and preferably 20 percent of potential earnings. This guideline is designed to simplify the determination and administration of benefits.

4. The wage-loss benefits should be a high percent of compensable wage loss, but not so high as to become a disincentive. The replacement rate should be 75 to 85 percent of any earnings losses in excess of the minimum required wage loss of 20 percent of potential earnings. A replacement rate in excess of this raises a serious work disincentive problem.

---

15. Additional specific factors that would have to be considered in formulating a hybrid system are discussed in Berkowitz, Burton, and Vroman (1979), chap. 18.

5. The earliest possible date of eligibility for wage-loss benefits is immediately after the scheduled benefits expire. The actual payment of wage-loss benefits must commence within a reasonable time period after the initial eligibility date in order to increase the probability that any earnings shortfall was caused by the work-related injury or disease. Possible cutoff dates for commencing wage-loss benefits would be two years after the date of maximum improvement for less serious injuries and three years after the date of maximum improvement for more serious injuries. These dates could then be extended as experience under the program accrues. The cutoff dates should alleviate the problems of cases being open for long periods and of determining the cause of the disability.

6. A worker who experiences continuing compensable wage loss should continue to receive benefits until normal retirement age. This should overcome the equity problems associated with the current *ex ante* approach to permanent disability benefits found in most state workers' compensation programs.

7. The potential earnings of a worker should be adjusted through time in order to reflect changes in the wages that the worker would have earned. These adjustments should be made by using a formula, such as adjusting each worker's wage in accordance with changes in the state's average weekly wage, rather than by making individual estimates on the basis of the worker's personal characteristics.[16] The latter procedure would be a source of considerable litigation. The purpose of adjusting the potential earnings is to provide more realistic estimates of the wage loss caused by the work-related injury or disease.

The intent of the hybrid system is to provide most cases only type I benefits, which, because of the emphasis on objective factors to produce a standard or modified rating, should reduce the amount of litigation in most states. This should help improve the efficiency of the program without adversely affecting adequacy or equity. The type II benefits would be used in those unusual cases where the type I bene-

16. A more complicated formula could adjust the worker's pre-injury wage by subsequent earnings developments in his industry or occupation, rather than using the state's average weekly wage. The crucial element is that the projections be done by a well-defined formula that does not permit litigation about potential earnings on a case-by-case basis.

fits were seriously deficient, thus improving the equity of the program. The hybrid system makes no pretense of providing complete equity in the sense of precisely matching benefits to wage loss, but this is in any case an illusory goal that would elude even a pure wage-loss system (which would also be a relatively inefficient system since the type of difficult decisions required in the type II benefits in the hybrid system would have to be made in all the cases in the pure wage-loss system). The hybrid system has as its virtue the provision of a safety valve for the worker who has unusually adverse experience.[17]

The Wage-Loss Concept in Florida

Among the most widely discussed recent developments in workers' compensation is the 1979 Florida legislation that incorporated the wage-loss concept into the permanent partial disability benefits there. The Florida legislation provides impairment benefits that are paid to workers with certain types of permanent impairments (amputations, loss of 80 percent or more of vision, or serious head or facial disfigurements) but are not paid to workers with other types of permanent impairments (such as total or partial loss of use of a body member). The impairment benefits were $50 for each percent of permanent impairment for 1 to 50 percent ratings, and $100 for each percent over 50 percent. Thus a worker with a 60 percent impairment rating would have received $3,500 under the 1979 law. These amounts were increased effective May 1, 1982, to $250 for each percent of permanent impairment for 1 to 10 percent ratings, and $500 for each percent over 10 percent. Thus a worker with a 60 percent impairment rating now receives $27,500. These impairment benefits can be paid in a lump sum as of the date of maximum medical improvement.

Workers with permanent impairments are also eligible for wage-loss benefits as of the date of maximum medical improvement if sufficient wage loss occurs. Wage-loss benefits are paid to workers

17. The hybrid system, while having only limited counterparts in actual practice, does have an intellectual history that provides some reassurance for those concerned about sharp breaks in tradition. Perhaps the most compelling precedent is the Report of the Permanent Partial Disabilities Committee of the IAIABC (Reid 1966). Although the particulars of the Reid committee report differ from the hybrid proposal in this section, its purposes and approaches are similar.

with at least a 15 percent loss of earnings. Benefits are 95 percent of the earnings losses in excess of the 15 percent threshold. The maximum wage-loss benefit is the lesser of 66.66 percent of the worker's pre-injury wage or 100 percent of the state's average weekly wage. There is, in effect, a 3 percent per year escalation in potential earnings used to calculate wage loss (this figure increased to 5 percent a year for injuries that occurred after July 1, 1980).

The maximum duration for the wage-loss benefits is 350 weeks (increased to 525 weeks for injuries that occurred after July 1, 1980) or age sixty-five (if the worker is eligible for social security benefits), whichever occurs first. Workers lose their eligibility for additional wage-loss benefits if they do not experience at least three consecutive months of compensable wage loss in each two-year period.

The 1979 Florida legislation thus establishes two types of benefits for workers with permanent consequences of their work-related injuries or diseases. Impairments benefits are paid to workers with the specified types of permanent impairments. Wage-loss benefits are paid to workers with wage losses that meet the statutory prerequisite. An individual worker may be eligible for impairment benefits, or wage loss benefits, or both types of benefits, or neither type of benefit — depending upon the exact nature of the permanent impairment and the timing and amount of the earnings losses.

The Disincentive Problem with the Florida Wage-Loss Benefits. One of the disincentive issues in workers' compensation concerns the relationship between the amount of benefits and the amount of lost earnings. The replacement rate can be defined as the amount of benefits divided by the amount of wage loss caused by the work-related injury or disease. If the replacement rate is too high, there is a disincentive for workers to return to work.

Table 2.2 analyzes the replacement rates for the wage-loss benefits under the 1979 Florida legislation. In 1979, the gross average weekly wage, that is, wages before deductions for taxes and other items, was $218.22 for Florida workers. The net average weekly wage, that is, gross wages minus deductions for federal income and social security taxes, was $193.51 for a married Florida worker with three dependents. The relationship between gross and net wages changes because of the progressive nature of the federal income tax.

TABLE 2.2
Wage Loss and Benefits
with 15-95 Wage-Loss Benefits,
Florida, 1979

	No injury	33.33% Gross Wage Loss	66.67% Gross Wage Loss	100% Gross Wage Loss
Gross wage[1]	$218.22	$145.48	$ 72.74	$ 0.00
Net wage[2]	193.51	139.27	71.92	0.00
Gross wage loss	...	72.74	145.48	218.22
Net wage loss	...	54.24	121.59	193.51
Compensable wage loss[3]	...	40.01	112.75	185.49
Benefit[4]	...	38.01	107.11	145.48[5]
Benefit as % of gross wage loss	...	52.3%	73.6%	66.6%
Benefit as % of net wage loss	...	70.1%	88.1%	75.2%

Notes: 1. Gross wage before injury is 100 percent of the average weekly wage earned in 1979 by all Florida employees covered by the Unemployment Insurance Act.

2. Net wage is spendable average weekly earnings using 1979 annual average formula for a married worker with three dependents found in Seifert 1981.

3. Compensable wage loss is gross wage loss minus 15 percent of gross wage before injury. This is the actual 1979 law in Florida.

4. Benefit is 95 percent of compensable wage loss. The maximum benefit is the lesser of 66.66 percent of the worker's pre-injury wage ($145.48 in this example) or 100 percent of the state's average weekly wage, which was defined as $195 for 1979 (effective August 1, 1979) in the Florida statute.

5. Benefit restricted to 66.66 percent of the worker's pre-injury wage.

For a worker who experiences a 33.33 percent loss of wages, the loss of gross wages is $72.74 and the loss of net wages is $54.24. Using the 15-95 benefit formula in the 1979 Florida legislation (see table 2.2, notes 3 and 4), the wage loss benefit received by the worker is $38.01, which represents 52.3 percent of the gross wage loss and 70.1 percent of the net wage loss. For a worker who loses 66.66 percent of pre-injury wages, the replacement rates are 73.6 percent of gross wage loss and 88.1 percent of net wage loss. Because of the maximum benefit amounts provided by the 1979 Florida legislation, the worker who has a total loss of income has replacement rates of 66.66 percent of gross wage loss and 75.2 percent of net wage loss.

The replacement rate data in table 2.2 do not appear too far out of line with the traditional notions that workers' compensation

TABLE 2.3
Marginal Replacement Rates with 15-95 Wage-Loss Benefits

	Move from 100% to 66.67% Gross Wage Loss	Move from 66.67% to 33.33% Gross Wage Loss	Move from 33.33% to 0% Gross Wage Loss
Change in benefits	$38.37	$69.10	$38.01
Change in gross wage	72.74	72.74	72.74
Change in net wage	71.92	67.35	54.24
Marginal replacement rate *G* (change in benefits ÷ change in gross wage)	52.7%	95.0%	52.3%
Marginal replacement rate *N* (change in benefits ÷ change in net wage)	53.4%	102.6%	70.1%

benefits should replace 66.66 percent of gross wage loss or 80 percent of spendable earnings loss. (National Commission 1972, pp. 56–57). The replacement rates for workers with a 66.66 percent loss of wages, however, provide a clue that problems may be lurking in the 15-95 wage-loss formula.

Another disincentive issue concerns the relationship between reductions in benefits that result from additional earnings. The marginal replacement rate can be defined as the change in benefits divided by the change in earnings that result from additional hours of work by an employee. If the marginal replacement rate is too high there is a disincentive for workers to return to work. To take an extreme example, if the marginal replacement rate is 100 percent and a $100 increase in earnings leads to a $100 reduction in benefits, only a devotee of the work ethic is likely to seek additional employment.

The disincentive issue associated with the marginal replacement rates included in the 1979 Florida legislation can be analyzed with the data in table 2.3. The table uses the information from table 2.2 for a Florida worker earning the state average weekly wage and in effect asks what happens to a worker who returns to work in stages. If the worker starts from a position of no earnings and goes back to work one-third time, his or her benefits will drop $38.37 (from $145.48 to $107.11), resulting in a marginal replacement rate for gross earnings of 52.7 percent and a marginal replacement rate for net earnings of

53.4 percent. If, as the next step, the worker increases work from one-third time to two-thirds time, benefits will drop $69.10 (from $107.11 to $38.01), resulting in a marginal replacement rate for gross earnings of 95.0 percent and a marginal replacement rate of 102.6 percent for net earnings. Finally, if the worker now expands the amount of work time from two-thirds to full time, the benefits will drop $38.01, resulting in a marginal replacement rate for gross earnings of 52.3 percent and a marginal replacement rate for net earnings of 70.1 percent.

There surely is a disincentive for the worker facing these choices. Several observations and qualifications are necessary, however. First, the marginal replacement rates depend on the extent to which the worker has already returned to work and the increment of full-time work he or she is contemplating. For example, a worker who has gone back to work one-third time and is considering a jump to full-time employment faces a marginal replacement rate that is the average of those shown in the second and third columns of table 2.3. Second, the marginal replacement rates depend on the level of the worker's pre-injury wage compared to the state's average weekly wage. For example, a worker earning twice the state's average weekly wage will have lower average and marginal replacement rates when sustaining total loss of wages because of the limit on wage-loss benefits to 100 percent of the state's average weekly wage. Third, the net wage loss and the replacement rates for net wage loss depend on each worker's tax status. The examples in tables 2.2 and 2.3 assume a worker with three dependents, and different numbers of dependents will change the net wage amounts. Fourth, the disincentive discussion implicitly assumes that workers have choices about the amount of work they can provide. To the extent employers, carriers, or the workers' compensation agency dictate the work schedule, the disincentive issue is irrelevant.

Even with these qualifications, the disincentive issue looms large in the Florida wage-loss benefits. Many workers with permanent impairments experience a long period of recovery while they are increasing their work hours from zero to full time; some workers with serious injuries never do return to work full-time. Vroman (1973) has provided some evidence demonstrating this recovery process among Florida workers with work-related injuries. For these workers, the high marginal replacement rates shown in table 2.3 are a deterrent to recovery unless the unrealistic assumption is made that workers have no control over the amount of time they work.[18]

The magnitude of the disincentive problem in the 1979 Florida legislation is not inherent in the wage-loss approach. The original proposal for wage loss benefits in Florida used a 20-80 formula instead of the 15-95 formula ultimately adopted.[19] The 20-80 plan is analyzed in tables 2.4 and 2.5, which show the highest marginal replacement rate for net wage loss is 86.4 percent. Even this figure ought not induce euphoria on the disincentive issue, but in comparison to the 15-95 plan, the 20-80 plan seems much preferable in terms of marginal replacement rates.

A possible objection to the 20-80 plan is that the relatively high threshold (20 percent loss of earnings) causes the average replacement rate to be too low for workers with modest losses of earnings. The average Florida worker with 33.33 percent loss of wages receives only $23.28 of benefits under the 20-80 plan (table 2.4), compared with $38.01 under the 15-95 formula actually adopted in Florida (table 2.2).

The problem with inadequate *average* replacement rates can be solved without relying on excessive *marginal* replacement rates. The key to the solution is to lower the threshold figure used to define compensable wage loss. Tables 2.6 and 2.7 illustrate a 10-80 plan. The average replacement rate for a worker with a 33.33 percent loss of earnings is higher under the 10-80 plan than under the 15-95 plan (56.0 percent of gross wage loss in table 2.6 versus 52.3 percent in table 2.2). The marginal replacement rate in the 10-80 plan, however, never approaches the excessive rate in the 15-95 plan (80.0 percent of gross wages in table 2.7 versus 95.0 percent in table 2.3). The lesson is that the generosity of a wage-loss plan should be determined by varying the threshold figure used to define compensable wage loss, rather than by varying the figure used to determine benefits as a percentage of compensable wage loss.

---

18. Several other chapters in this volume provide evidence that higher benefits induce greater amounts of work disability. See, however, the New York results later in the present chapter.

19. The 20-80 formula (then known as the 80-80 formula) was proposed by Wayne Vroman and John Burton to the Florida Workers' Compensation Advisory Committee in January 1978.

TABLE 2.4
Wage Loss and Benefits with 20-80 Wage-Loss Benefits, Florida, 1979

	No injury	33.33% Gross Wage Loss	66.67% Gross Wage Loss	100% Gross Wage Loss
Gross wage[1]	$218.22	$145.48	$ 72.74	0.00
Net wage[2]	193.51	139.27	71.92	0.00
Gross wage loss	...	72.74	145.48	218.22
Net wage loss	...	54.24	121.59	193.51
Compensable wage loss[3]	...	29.10	101.84	174.58
Benefit[4]	...	23.28	81.47	139.66
Benefit as % of gross wage loss	...	32.0%	56.0%	64.0%
Benefit as % of net wage loss	...	42.9%	67.0%	72.2%

Notes: 1. Gross wage before injury is 100 percent of the average weekly wage earned in 1979 by all Florida employees covered by the Unemployment Insurance Act.

2. Net wage is spendable average weekly earnings using 1979 annual average formula for a married worker with three dependents found in Seifert 1981.

3. Compensable wage loss is gross wage loss minus 20 percent of gross wage before injury.

4. Benefit is 80 percent of compensable wage loss. The maximum benefit is 100 percent of the state's average weekly wage, which was defined as $195.00 for Florida in 1979.

TABLE 2.5
Marginal Replacement Rates with 20-80 Wage-Loss Benefits

	Move from 100% to 66.67% Gross Wage Loss	Move from 66.67% to 33.33% Gross Wage Loss	Move from 33.33% to 0% Gross Wage Loss
Change in benefits	$58.19	$58.19	$23.28
Change in gross wage	72.74	72.74	72.74
Change in net wage	71.92	67.35	54.24
Marginal replacement rate *G* (change in benefits ÷ change in gross wage)	80.0%	80.0%	32.6%
Marginal replacement rate *N* (change in benefits ÷ change in net wage)	80.9%	86.4%	42.9%

Evaluation of the Florida Wage-Loss Benefits. The 1979 Florida legislation can be evaluated in terms of the seven guidelines for a hybrid system provided earlier.[20] The record is mixed. The Florida legislation substantially meets the third guideline, which requires that wage-loss benefits be only paid to workers with at least modest lost earnings (the requisite in Florida that there be at least a 15 percent loss before any wage-loss benefits are paid is a minimal figure for the threshold guideline). Likewise, the Florida legislation meets the fifth guideline, pertaining to a reasonable period after the date of maximum medical improvement when wage loss must first be established in order to be eligible for the wage-loss benefits. The Florida law also partially complies with the seventh guideline, which requires the potential earnings of workers to be adjusted through time to reflect changes in wages. The annual escalation of 3 percent (or 5 percent) in the procedure used to calculate wage loss is rather low in light of the increases in prices and wages in recent years, but at least the Florida law provides a modest contribution to protecting workers against inflation.

The Florida legislation fails to meet four of the guidelines for a hybrid system. The first guideline, requiring a lag between the date of maximum medical improvement and the initial date of eligibility for wage-loss benefits, is clearly violated since workers are immediately eligible for benefits. Many workers will begin to receive benefits in the rehabilitation phase of their post-injury recovery period and are likely to face a serious disincentive. The disincentive problem is aggravated because 95 percent of earnings losses are replaced, a clear violation of the fourth guideline. This is especially troublesome since the benefits in Florida are tax-free. With a 95 percent replacement rate, some workers will actually be worse off by increasing the amount of time they work.

The Florida law also violates the sixth guideline, which suggests that workers who experience continuing wage loss should continue to receive benefits until their normal retirement age. At most,

---

20. The 1979 Florida legislation represents neither a pure wage-loss system nor a hybrid system as those systems have been defined. Nonetheless, I apply the seven general guidelines for a hybrid system to the Florida legislation because they provide criteria that can be used to evaluate any state's permanent partial disability benefits. States that use an *ex ante* system would be found seriously defective in terms of these seven guidelines.

TABLE 2.6
Wage Loss and Benefits with 10-80 Wage-Loss Benefits, Florida, 1979

	No injury	33.33% Gross Wage Loss	66.67% Gross Wage Loss	100% Gross Wage Loss
Gross wage[1]	$218.22	$145.48	$ 72.74	0.00
Net wage[2]	193.51	139.27	71.92	0.00
Gross wage loss	...	72.74	145.48	218.22
Net wage loss	...	54.24	121.59	193.51
Compensable wage loss[3]	...	50.92	123.66	196.40
Benefit[4]	...	40.74	98.93	157.12
Benefit as % of gross wage loss	...	56.0%	68.0%	72.0%
Benefit as % of net wage loss	...	75.1	81.4	81.2

Notes: 1. Gross wage before injury is 100 percent of the average weekly wage earned in 1979 by all Florida employees covered by the Unemployment Insurance Act.

2. Net wage is spendable average weekly earnings using 1979 annual average formula for a married worker with three dependents found in Seifert 1981.

3. Compensable wage loss is gross wage loss minus 10 percent of gross wage before injury.

4. Benefit is 80 percent of compensable wage loss. The maximum benefit is 100 percent of the state's average weekly wage, which was defined as $195.00 for Florida in 1979.

TABLE 2.7
Marginal Replacement Rates with 10-80 Wage-Loss Benefits

	Move from 100% to 66.67% Gross Wage Loss	Move from 66.67% to 33.33% Gross Wage Loss	Move from 33.33% to 0% Gross Wage Loss
Change in benefits	$58.19	$58.19	$40.74
Change in gross wage	72.74	72.74	72.74
Change in net wage	71.92	67.35	54.24
Marginal replacement rate *G* (change in benefits ÷ change in gross wage)	80.0%	80.0%	56.0%
Marginal replacement rate *N* (change in benefits ÷ change in net wage)	80.9	86.4	75.1

the Florida law will provide wage-loss benefits for slightly more than ten years. This is not a trivial period, however, compared to the maximum duration in many jurisdictions, and may be justified at least partially because of the newness of the Florida law.

The final violation of the guidelines offered in the previous section is that the Florida law does not provide some benefits to all workers with a permanent impairment, as required by the second guideline. Indeed, to take an extreme example, a worker who suffers a 100 percent loss of use of an arm and returns to the old job with no wage loss will receive no permanent disability benefits. It is hard to justify a system that will provide permanent impairment benefits to a worker who experiences the amputation of a fourth finger, but no impairment benefits for a worker who experiences an injury that leaves him or her with an arm that is useless.

An overall evaluation of the 1979 Florida legislation dealing with permanent disabilities is obviously difficult and dependent on subjective values. Using the seven guidelines in the previous section, the law seems defective. On one hand, the impairment benefits are much too restrictive, particularly in terms of the number of workers who are eligible. On the other hand, the wage-loss benefits seems unduly generous because an extremely high proportion of wage loss will be replaced for workers who qualify for these benefits, thus virtually ensuring a disincentive problem. Further, the disincentive problem is aggravated by the fact that the benefits are available immediately after the date of maximum medical improvement, when many workers are in a critical phase of the rehabilitation process.

Programs providing for permanent partial disability benefits in workers' compensation are in a period of ferment that can be traced in large part to the 1979 adoption of the wage-loss approach in Florida. The Florida legislation was a product of political compromises and clearly has merits as well as deficiencies. The challenge for reform of permanent partial disability benefits in the next few years is whether the Florida approach can be transferred to other jurisdictions with some of the merits intact while the defects are deleted.

Permanent Partial Disability Benefits in New York

The New York workers' compensation law provides some additional guidance for improving delivery of permanent partial disability benefits in general and wage-loss benefits in particular (Burton, Larson, and Moran 1980).

Scheduled Benefits. The scheduled permanent partial disability benefits in New York are a typical example of *ex ante* benefits and are representative of those found in most jurisdictions. A list of body members is included in the statute with corresponding durations of benefits.[21] For example, a worker who suffers physical loss or loss of use of an arm receives 312 weeks of benefits, a leg, 288 weeks, and so on, down to 15 weeks for loss of a fourth finger. For partial loss or loss of use, compensation is paid for a period of time proportionate to the degree of loss. The weekly benefits are 66.66 percent of the worker's pre-injury wage, subject to a maximum of $105 per week. A hearing to determine the duration of the permanent disability benefits is scheduled as soon as the permanent consequences of the injury can be rated and, with the exception of the major member benefits described below, the duration is not affected by the existence of actual wage loss.

Nonscheduled Benefits. The New York nonscheduled permanent partial disability benefits are 66.66 percent of the difference between the wages earned before the injury and the wages the worker is earning, or is able to earn, after the injury. The benefits are subject to a maximum of $105 a week and a minimum of $20 a week (or actual wages if less than the minimum). The nonscheduled benefits are paid for the full period of disability, which can mean lifetime benefits. The Workers' Compensation Board can modify the amount of the benefits upon a showing of change in the worker's condition or earning capacity.

Several aspects of these nonscheduled benefits in New York should be mentioned. A worker with a nonscheduled permanent impairment receives no benefits unless there is an actual loss of earnings. In 1977 (the latest year for which data are available), 8.5 percent of nonscheduled cases were closed with no permanent disability benefits paid because there was no current loss of earnings. These cases are eligible to be reopened for a period of eighteen years from the date of the original injury or eight years from the last payment of compensa-

21. The duration of the scheduled benefits is reduced (within limits) by the duration of temporary disability payments.

tion, whichever is later. Nonscheduled cases can be closed with lump-sum awards, but these are subject to stringent preconditions and, unlike the lump-sum procedure used in most states, the New York lump-sum cases can be reopened upon a showing of change in medical condition. In any event, the lump-sum awards represent a declining share of nonscheduled cases — down from 45 percent in 1970 to 26 percent in 1977. The clear majority (65 percent in 1977) of nonscheduled cases in New York are closed administratively with payments continuing because the worker has demonstrated continuing actual wage loss.

There are additional aspects of the nonscheduled benefits that complicate the categorization of the New York approach, and data on the administration of the approach are inadequate.[22] The essential point, however, is that the New York nonscheduled benefits are operationally based on actual loss of earnings and paid on an *ex post* basis.[23] Indeed, using the criteria for identifying a pure wage-loss approach, the New York nonscheduled benefits are probably the most pristine version of the approach found in the country. What makes the New York wage-loss approach purest among the semipure is that benefits can be paid as long as wage loss continues (other states limit duration), and compromise and release agreements or other devices to limit the potential duration of wage-loss benefits are uncommon in New York (Michigan makes extensive use of redemption settlements). The reason the New York wage-loss benefits are not so well known as those in the other jurisdictions is that the New York schedule encompasses a high proportion of all injuries with permanent consequences, leaving only a relatively few nonscheduled cases. In contrast, the scope of the Michigan and Pennsylvania schedules is quite limited, leaving most cases to be handled by the wage-loss approach, and the 1979 Florida statute makes wage-loss benefits potentially available to all workers with permanent disabilities (including those

---

22. There is inadequate information on reopened cases.

23. The nonscheduled benefits that nominally are due can be reduced if the Workers' Compensation Board determines that there is a smaller loss of earning capacity. This limitation, similar to a provision in the 1979 Florida legislation, is almost inevitable even in a pure wage-loss aproach. An employer retains the right to demonstrate that the worker's post-injury earnings are artificially depressed by the worker's voluntary restriction of labor supply.

workers who also receive scheduled or impairment benefits). Although the New York nonscheduled benefits are relatively unknown, the experience with the approach is enlightening, and will be reviewed in the next section.

Major Member Continuing Disability Benefits. Section 15(3)(v) of the New York workers' compensation law provides that in cases involving 50 percent or more loss or loss of use of an arm, leg, hand, or foot, benefits are paid for the duration of the scheduled benefits, and, if actual wage loss continues beyond the scheduled duration, additional benefits are paid. The weekly benefit during the scheduled period is 66.66 percent of the worker's pre-injury wage (subject to the $105 weekly maximum), and the weekly benefit after the scheduled period is 66.66 percent of the difference between the worker's pre-injury wage and the worker's actual earnings after the scheduled benefits have expired (again subject to the $105 weekly maximum). The provision thus shifts benefits from an *ex ante,* proxy approach to an *ex post,* actual wage-loss approach and serves as a prototype of the hybrid system described earlier. Obviously, the version of the hybrid system encompassed in Section 15(3)(v) varies in a number of particulars from the hybrid system proposed here, but the essential attribute of the approach — starting with presumed disability benefits and then shifting to actual wage-loss benefits — is present.

Developments in Permanent Partial Disability Cases in New York

New York makes use of an *ex ante* approach, a wage-loss (or *ex post*) approach, and a hybrid system, making the Empire State sui generis.

In New York, the relative importance of scheduled and nonscheduled disabilities can be shown by statistics derived from the annual reports on closed compensated cases, but unfortunately the data on the major member continuing disability benefits (the hybrid approach) are too sparse to allow an historical review. The proportion of all compensated cases closed with an award for either scheduled or nonscheduled permanent partial disability increased from 9.7 percent in 1916 to 36.9 percent in 1955 to 41.4 percent in 1965. Since then the percentage has fluctuated in the 39 to 41 percent range, with scheduled awards accounting for 35 to 39 percent and nonscheduled awards representing 2 to 4 percent of all cases (Burton, Larson, and Moran 1980, tables 9.2 and 9.4).

While the proportion of all cases accounted for by permanent partial disability cases has not varied much in recent decades, the costs of these benefits have shown considerable movement. From 1970 to 1979, for example, the percentage of all compensation costs accounted for by scheduled awards declined from 35.7 to 23.9 percent, while the nonscheduled awards increased their share from 34.5 percent to 54.6 percent of all compensation costs. The nonscheduled cost developments are particularly striking: in 1979 the nonscheduled awards represented 3.9 percent of the cases and over half of the dollars (Burton, Larson, and Moran 1980, tables 9.2 and 9.4).

I developed a model to explain the variations through time in the number, the average cost, and the total cost of the New York cases closed each year that paid permanent partial disability scheduled or nonscheduled benefits. The model is adapted from the approach used by Butler and Worrall (1982).[24]

The dependent variables measuring number of cases closed are the counts as published by the New York Workers' Compensation Board, divided by the number (in thousands) of New York employees.[25] The dependent variables involving costs are average cost per case and total cost per thousand employees, both in constant dollars (1967=100).

The Butler-Worrall model used simultaneous equations to explain wages, hours, and injury rates. I use simultaneous equations involving wages, hours, and (in turn) number of cases, average cost, and total cost. Butler and Worrall analyzed three types of cases — temporary total, permanent partial major, and permanent partial minor — and constructed expected benefits for each type. I analyze two types of cases — scheduled permanent partial and nonscheduled permanent partial. For institutional reasons, however, the New York expected benefits are identical for both types of permanent partial

---

24. The adaptations made for this study are discussed in an appendix, which is available from the author.

25. Butler and Worrall (1982, p.4) use the number of claims filed per thousand employees as their measure of the injury rate. The New York data used in this study measure the number of cases closed per year, which has been divided by the number (in thousands) of New York employees. Because of the lag between date of filing and date of closing, all regressions in this study lag the independent variables by two years.

TABLE 2.8
Number and Cost of
Permanent Partial Disability Cases:
Expected Signs on
Selected Independent Variables

Independent Variables	Number of Cases		Average Cost		Total Cost	
	Scheduled	Nonscheduled	Scheduled	Nonscheduled	Scheduled	Nonscheduled
Temporary total benefit	+	+	+	+	+	+
Permanent partial benefit	+	+	+	+	+	+
Unemployment rate	+	++	+	++	+	++

Notes: Number of cases, average cost, and total cost are dependent variables.
+ expected sign is positive.
++ absolute size should be larger than the absolute size of +.

cases. The Butler-Worrall data are annual observations for the period 1972–78 for thirty-five jurisdictions. I use New York annual data for the period 1959–79, the only years for which comparable New York data are available. As a result, some of their variables are not used because data are unavailable or the variables were inappropriate. Finally, Butler and Worrall do not use the unemployment rate for their injury rate equations. I use the unemployment rate as an explanatory variable for the number of cases closed, the average cost, and the total cost equations.

The independent variables of particular interest are the temporary total benefit, the permanent partial benefit, and the unemployment rate. The expected signs for these variables are shown in table 2.8. The expected coefficients of the benefit variables are related to an institutional feature of workers' compensation. Almost all cases involving cash benefits begin with temporary total benefits. Some of these cases subsequently receive permanent partial benefits. If this occurs, the case is classified as a permanent partial disability case (even though the case began as a temporary total case), and the data show all the benefits in the case as payments for permanent partial disability. Given this institutional feature, Butler and Worrall argue that "the size of the temporary total payments should have a positive effect on the number of permanent partial injuries filed'' since the temporary total benefits bring some new cases into the program, and some of these cases end up with permanent partial awards (1982, p. 18). Moreover, Butler and Worrall assert that the major and minor permanent partial disability benefits are interrelated: increases in major benefits increase major claims and increases in minor benefits increase minor claims. In general, their empirical work supports these hypothesized relationships.

Consistent with the Butler and Worrall analysis, table 2.8 shows an expected positive coefficient on temporary total benefits for the numbers of the two types of permanent partial disability cases. I have only one benefit variable for both types of permanent partial cases and so cannot show how shifts in the relative benefits lead to changes in the proportion of scheduled and nonscheduled cases. An increase in permanent partial benefits, however, should increase the numbers of both types of permanent partial cases.

The unemployment rate is also included as a variable in the equations for number of cases closed. Since the extent of work disability for a given worker depends not only on the extent of functional limitations but on such other influences as labor market conditions, the number of cases closed in a year should be positively associated with the unemployment rate. But since scheduled permanent partial disability benefits are paid regardless of wage loss, while nonscheduled permanent partial benefits are only paid if the worker experiences an actual loss of earnings, the coefficient on the unemployment rate is expected to be larger in the regression for nonscheduled benefits than in the regression for scheduled benefits.

The regression results for the number of cases are presented in table 2.9. The coefficients on temporary total benefits are insignificant in both regressions, while the coefficient is significant for permanent partial benefits only in the regression for nonscheduled permanent partial cases. The sign, however, is negative, contrary to expectations. The coefficient for the unemployment rate is insignificant in the regression for scheduled permanent partial cases and is significant with the expected positive sign in the regression for nonscheduled permanent partial cases. The table indicates that a 1 percent increase in the unemployment rate is associated with a 0.323

TABLE 2.9
Number and Cost of
Permanent Partial Disability Cases,
New York, 1959–79

Independent Variables	Number of Cases		Average Cost		Total Cost	
	Scheduled	Nonscheduled	Scheduled	Nonscheduled	Scheduled	Nonscheduled
Temporary total benefit	−.520 (.618)	.395 (.549)	.560 (1.13)	.514 (.956)	−.372 (.511)	1.27 (1.13)
Permanent partial benefit	.368 (.952)	−1.68 (5.07)***	.454 (1.69)	−.263 (.905)	.836 (2.50)**	−1.95 (3.79)***
Unemployment rate	−.073 (.513)	.323 (2.65)**	−.044 (.587)	.167 (2.07)*	−.204 (1.65)	.567 (2.98)***
R^2	.315	.833	.582	.946	.690	.914
F	1.07	11.7***	4.17**	53.0***	5.18***	24.9***
Durbin-Watson	1.44	2.39	.984	1.45	2.07	1.85

Notes: Number of cases, average cost, and total cost are dependent variables.

All variables shown in logs; the benefit variables are dollars in real values.

These variables were included with other independent variables in two-stage least squares regressions that are reported in full in the appendix, where the variables are defined, values for the dependent variables given, and sources provided. The appendix is available from the author.

Statistical tests: *t*-ratios in parentheses.

Significance levels: *(0.10) **(0.05) ***(0.01)

percent increase in the number of nonscheduled permanent partial cases.[26]

The Butler and Worrall model did not consider determinants of the average cost per case and the total cost per thousand employees, but the model can be expanded to include these dependent variables. Higher temporary total benefits can be expected to increase the average costs and the total costs of both types of permanent partial cases, as shown in table 2.8.[27] Higher permanent partial benefits should also increase the average and total costs of the scheduled and nonscheduled cases. The unemployment rate should be positively associated

26. The variables are entered in log form, so coefficients can be interpreted as elasticities.

27. Total cost per thousand employees is the product of the number of cases per thousand employees and the average cost per case; since the numbers and averages are expected to be positive, the totals should also be positive. Similar reasoning explains the other expected signs in table 2.8.

with average costs and total costs for both types of cases.[28] Again, the coefficients on the unemployment rate are expected to be larger in the regressions for the nonscheduled benefits than in the regression for scheduled benefits.

The regression results for the average costs of closed cases are included in table 2.9. In the regressions for scheduled permanent partial benefits, both benefit coefficients have the expected positive sign, but neither is significant. For nonscheduled cases, the signs on the benefit coefficients are mixed and insignificant. The only significant coefficient for the unemployment rate in the average cost regressions is for the nonscheduled cases; a 1 percent increase in the unemployment rate is associated with a 0.167 percent increase in average cost.

For the total costs of closed cases, two of the coefficients for benefits are insignificant; the two significant coefficients suggest that higher permanent partial disability benefits increase the total cost of scheduled cases and decrease the cost of nonscheduled cases. The only significant coefficient on the unemployment rate indicates that higher unemployment is associated with a higher total cost for nonscheduled permanent partial cases.

The results in table 2.9 taken as a whole do not provide much support for the aspects of the model summarized in table 2.8. The significant coefficients on benefits have the wrong sign in two of three instances and are insignificant nine times. The unemployment rate coefficients are insignificant three times, although all three significant coefficients have the expected sign. Probably the most compelling results involve the statistically significant relationships between the unemployment rate and nonscheduled permanent partial cases: a deteriorating labor market is associated with more nonscheduled cases and with higher average and total costs for these cases.

To illustrate the possible importance of the relationship between labor market conditions and nonscheduled permanent partial benefits, consider these developments between 1970 and 1979. The variables in table 2.9 were entered in log form, so the coefficients

28. The higher unemployment rate should increase the duration of temporary total disability benefits in both types of permanent partial cases and also the duration of nonscheduled permanent partial disability benefits, which are paid only when actual wage loss occurs.

indicate the percentage increase in the dependent variable associated with a 1 percent increase in the unemployment rate. Over this period, nonscheduled cases were up 54.6 percent (from 0.560 per thousand employees to 0.866 per thousand), while the unemployment rate (lagged two years) was up 96.0 percent (from 2.5 to 4.9). Given the elasticity on the unemployment rate of 0.323, the increase in the unemployment rate would have produced a 31.0 percent increase in the number of nonscheduled cases. Likewise, the increase in the unemployment rate between 1970 and 1979 would have produced a 16.0 percent increase in the average (real) cost of nonscheduled cases, compared to the actual increase of 23.5 percent. Further, the unemployment rate increase would have increased total (real) costs per thousand employees by 54.4 percent, compared to the actual increase of 91.0 percent. Thus, over half the increases during the 1970s in all three measures of nonscheduled permanent partial benefits is associated with increases in the unemployment rate.

The results of applying the model to New York data must be viewed with caution, given the less than total consistency between predicted signs in table 2.8 and empirical results in table 2.9. But at the minimum, the statistical results suggest that scheduled and nonscheduled permanent partial benefits in New York are influenced by different factors, and that the number and cost of nonscheduled benefits are much more sensitive to labor market conditions. The differential results for scheduled and nonscheduled permanent partial benefits are consistent with expectations.

Conclusions

Several aspects of the preceding analysis merit emphasis. The dominant approach to compensating permanent partial disabilities relies on proxies for wage loss that are assessed before the period when the actual wage loss occurs. The data suggest that the *ex ante,* proxy approach does a poor job of matching benefits to actual wage loss and results in a serious equity problem. The results also suggest that the *ex ante* approach can provide benefits that on average are adequate, or inadequate, or more than adequate, and that the approach can be efficient or inefficient. In short, there is nothing inherent in the *ex ante,* proxy approach that precludes success on the adequacy and efficiency criteria, but I believe the approach can not achieve equity.

The wage-loss approach has received considerable attention in recent years because of the incorporation of a variant of the approach in the 1979 Florida legislation. The primary force behind the legislation probably was the conviction of Florida employers that before 1979 the law was unduly expensive because, as administered, the permanent partial benefits were too generous. The particular version of wage-loss benefits that Florida adopted apparently will reduce the employers' costs of workers' compensation, although additional time is necessary to confirm that tentative conclusion. But the lesson from New York's experience with wage-loss benefits is that the approach also can be expensive. Moreover, the New York data suggest that the cost of wage-loss benefits is more sensitive to labor market conditions than the cost of benefits paid on an *ex ante* basis. All of this suggests that wage-loss benefits are not inherently more or less expensive, or adequate, than the *ex ante* benefits they are likely to replace. The relative costs depend on the generosity of the *ex ante* benefits being replaced, the particular variant of wage-loss benefits adopted, and the condition of the labor market.

The added sensitivity of the cost of wage-loss benefits to labor market conditions is not necessarily a fault. Indeed, the results from New York suggest that the nonscheduled (wage-loss) benefits are more equitable than the scheduled benefits. There is no particular reason to expect that the actual adverse labor market experience of workers with nonscheduled benefits is more sensitive to higher unemployment rates than the experience of workers with scheduled benefits. It is rather that benefits in nonscheduled cases have more flexibility in terms of reflecting lost wages, which should result in more equity.

Although the wage-loss approach should improve the equity of benefits, it has its own faults, including disincentives for workers to return to work at a critical phase of the rehabilitation process and the absence of payments to workers with serious impairments but no wage loss. To overcome some of the problems inherent in the wage-loss approach and in the *ex ante* approach, I favor a hybrid approach that in effect combines in sequence the *ex ante* and wage-loss approaches.

When the seven guidelines offered for a hybrid system were used to evaluate the 1979 Florida wage-loss benefits, a number of deficiencies were noted. The wage-loss approach found in New York's nonscheduled permanent partial disability benefits would also fail to meet most of the seven guidelines, although the particulars of the Florida and New York score cards would differ.

The criticism of the wage-loss approaches in Florida and New York should not be misinterpreted. It is not a defense of the *ex ante* system that existed in Florida before 1979 and that exists in most states now. Rather, the criticism should be understood as a denouncement of both the *ex ante* and wage-loss approaches and as a paean to the hybrid approach. Of course, like the other two approaches, there can be considerable variations in design and administration of the hybrid approach. The New York variant of the hybrid approach — the major member continuing disability benefits — has deficiencies, such as too restrictive eligibility rules and underutilization that can be traced to poor administration. The deficiencies and possible remedies have been detailed elsewhere (Burton, Larson, and Moran 1980). The essential point is that a working example of a hybrid approach exists and can serve as a starting point for other states and allay concerns that the idea is an ivory-tower while-away.

3 · WAGE AND INJURY RATE RESPONSE TO SHIFTING LEVELS OF WORKERS' COMPENSATION

Richard J. Butler

As government at all levels seeks to further regulate and "protect" its citizenry, there has been an increasing reliance on microeconomic analysis of many of the more controversial and wide-ranging social welfare and insurance programs. Interestingly enough, one major public policy area, workers' compensation, has received very little attention. Virtually no empirical analysis has been conducted in this area despite the prominence in public policy discussions of the revision of the system of state laws known collectively as the workers' compensation laws.[1] Much of the debate centers around the recommendations of the National Commission on State Workmen's Compensation Laws, which urges a substantially larger federal role in unifying these laws with a change in the benefit structure that implies an extensive expansion of this form of social insurance. While these revisions may lead to the improvement in the income security of workers

Comments by John D. Worrall and members of the labor seminar at Cornell University are greatly appreciated, as is financial support from the Center for the Study of the American Political Economy. — R.J.B.

1. The article by Chelius and the manuscript of Dorsey and Walzer remain the only empirical work, to my knowledge, that examines the relationships among wages, injury rates, and the parameters of a workers' compensation system. Accidents are, of course, also affected by the level of safety effort by the firm as well as any safety rules it may establish. We abstract from these factors in the text, and estimate their effect in the empirical analysis by including a cyclical variable (*DVAL*), an intensity of work proxy (*DAY*), a secular trend (*TIME*), and industry-specific dummies.

that the commission intended, it is also clear that the changes — in their proposed form — are of such a magnitude that any unanticipated or secondary effects of the expansion are likely to have an important influence on the labor market. This paper analyzes two such important effects, the effect of a change in the structure of benefits on both injuries and wages.

Since the commission's 1972 report, several states have adopted its recommendations. Concomitant with the adoption of more liberal benefit standards, however, there has been a recent surge in the cost of the workers' compensation program. Not only have expenditures been increasing exponentially, but these expenditures as a percentage of the covered payroll have also been climbing rather dramatically.[2] These increases have been unanticipated, even though projections are made on the basis of a weighted extrapolation of benefits' costs based on earlier accident rates and recovery periods. It appears that the "more adequate" benefits may have affected workers' safety behavior, their propensity to file a workers' compensation claim, and time spent recuperating after an accident and thus have tended to increase the costs of the workers' compensation system. Despite the need for estimates of the elasticities of such behavior, little research has been done.

In this paper, I begin to fill this hiatus in empirical research by examining the effects of changing benefit levels on injury rates and the structure of wages. In the past when benefits have been liberalized, for example, cost projection has often proceeded on the implicit assumption that the claim filing frequency desired by the firm or the worker was unaffected by the size of the average benefit paid out to claimants. That is, there has often been a presumption of a completely inelastic response of claims to benefit changes. My empirical results indicate, however, that the response is not inelastic but rather positively related and, indeed, of such a magnitude both to warrant its consideration in future analysis of a program and to suggest the need for additional research. In addition I find that wages tend to decrease as benefits increase and to increase as claims rise. This is consistent with both the hypothesis that workers are aware of job risk and *ex post* injury payments, and the hypothesis that at least some of the change in claim rates reflects real changes in levels of risk induced by workers'

2. See table 1.1 in Chapter 1.

compensation (as opposed to a simple change in the propensity to report claims for any given level of risk).

Injuries and Benefits: Economic Theory

While most employees would not purposely have an accident, the theoretical model here is built on the notion that the worker is willing to face a greater degree of injury risk when the cost of having an accident is lower (or to take a longer convalescent period when the opportunity costs for doing so are lower)[3] or to file more claims for any specified number of accidents. While it is difficult to disentangle this latter reporting effect from the safety effect, using the claim data alone, the interaction of benefits, claims, *and* wages provides some information about the relative importance of the safety effect.

The pecuniary cost to the worker of receiving workers' compensation is relinquished wages, while gains are the indemnity benefits associated with a particular type of claim. The "injured" state then becomes relatively more attractive as benefits increase or as wages fall.[4] When employee effects are dominant, we expect to see

---

3. For the present, however, the discussion in the text centers on the probability (frequency) of accidents rather than the length of recovery, as this applies more directly to the estimates presented here.

4. We can translate these notions of opportunity cost and benefits directly into an empirical model by considering the indirect utility function of the ith state for the representative employee, $V^i(W^i,\phi^i)$, where the first argument in the indirect utility function is the marginal pay in the ith state and ϕ denotes a vector of observed (Z) and unobserved (ϵ) characteristics of individuals in the market. A linear approximation of the V^i function can then be written as

$$V^i(W^i,\phi^i)=\alpha_0^i + \alpha_1^i W^i + \alpha_2^i Z^i + \epsilon^i. \quad (3.1)$$

To put this in the context of the workers' compensation analysis, let $i = 1$ when there is no injury, and $i = 2$ when there is an injury. The wages (marginal pay) in this latter state will be the weekly benefit amounts, denoted by B. The relative attractiveness of the two states varies with value of V^1 and V^2, the injured state becoming more tolerable as $V^2 > V^1$ or when

$$\alpha_0^2-\alpha_0^1 + \alpha_1^2 B-\alpha_1^1 W + \alpha_2^2 Z^2-\alpha_2^1 Z^1 > \epsilon^1-\epsilon^2. \quad (3.2)$$

This forms the basis for the estimates with the expected positive sign on the coefficient of the wage and benefit variables:

$$(\alpha_1^1, \alpha_1^2 > 0).$$

That is, our expectation is that accidents increase, *ceteris paribus,* as benefits increase or wages decrease.

claims increase when benefits are liberalized or when wages fall, other things being the same.

But how likely is it that other things are the same? What about the other institutionally important aspects of workers' compensation? In our sample, the neglect of the nonpecuniary institutional aspect of workers' compensation is more apparent than real since our data come from a single state during a period of relatively little change in the workers' compensation system. Many of the system's parameters (duration of benefits, treatment of claims, underlying benefit parameters) remained relatively unchanged over time and so can be treated as constants (and captured in the intercept term). Such differences as the controversion of claims across industries, personnel practices, and training programs, which also affect the number and types of claims filed, are likewise captured by the vector of industry-specific dummy variables included in all the regressions.

Note that if employers' incentives to provide safer workplaces are little affected by changes in workers' compensation benefits *and* wages, when estimated from observable data these would have a structural interpretation (that is, they would mirror workers' responses to changes in the workers' compensation benefit structure). Obviously, however, to the extent that regressors also reflect firms' incentives, the estimates would be reduced-form in the sense that they compound both employers' and employees' incentives.

In a system such as workers' compensation, a change in benefits will not only affect the employees' incentives but it would also be expected to have an effect on the employers' incentives to provide safer workplaces. The firm, indirectly at least, finances workers' injury payments via insurance premiums, so that as the cost of the program increases, the firm's incentive to provide safer working conditions is also increased. This could mean that an increase in workers' compensation benefits will result not only in more adequate payments, but it may be possible that when employers' incentives are dominant, a workers' compensation program can provide incentives for safety as well.

The link between employers' costs and benefits is captured in the degree of experience rating. A fully (or perfectly) experience-rated firm bears *all* the costs of workers' compensation claims originated from that firm and hence has considerable incentive to reduce hazardous conditions (or alternatively perhaps, to fight more claims). A non-rated firm has little, if any, incentive to curtail claim costs. Since our sample consists of several small, highly competitive industries in South Carolina, experience rating is likely to be imperfect. The link between insurance costs and a particular firm's accident experience will be weak. For this reason, we anticipate that increased benefits will have a greater effect among employees than employers — so that a liberalization of the benefit structure will increase accident rates — and that we can isolate an employee effect. This means that one-tailed *t*-tests are appropriate in evaluating the wage and benefit coefficients.

South Carolina Industrial Data

The South Carolina Industrial Commission and the South Carolina Department of Labor provided extensive information on industrial accidents and on wages and employment, respectively, from which a cross-sectional sample of fifteen industries tracked over thirty-two years was gathered.[5] Both provided information for a fiscal year from July 1 to June 30. Most important, there are fifteen industries in any of the thirty-two years for which we have both information on their industrial characteristics as well as on injuries and compensation received under South Carolina's Workmen's Compensation laws. The industrial data came from the annual reports of the South Carolina Department of Labor, which compiled the data from mandatory annual reports that all industrial firms filed with the commission. This information has been aggregated to the industry level, and data for those industries reported in table 3.1 were matched with similar data reported by the South Carolina Industrial Commission. The latter recorded information for all claims and the total aggregate costs for cases that originated and closed during the respective fiscal year. This pooled cross section–time series will provide the bulk of the informa-

5. This leads to a sample size of 480. Our actual sample was slightly smaller, 468, as the industrial categories reported in the workers' compensation data changed slightly. A few industries were merged, and a few others no longer reported after 1969. Chow tests for structural shift on a full sample relative to the restricted 1940–69 sample indicated no structural shift in the injury rate regressions but some change in the wage equation. Since the latter change never resulted in any qualitative changes and did not affect any of the conclusions reached here, we report only results for the full sample.

TABLE 3.1
Factor Limit Calculations
for South Carolina

Industry	1950 No Tax	1950 Tax	1960 No Tax	1960 Tax	1970 No Tax	1970 Tax
Barrels	.926	.963	.862	.928	.582	.709
Brick	.729	.915	.738	.874	.439	.558
Chemicals	.769	.899	.643	.766	.473	.519
Clothing	.860	.874	.849	.874	.620	.718
Electricity	.675	.747	.553	.688	.373	.439
Fertilizer	.839	.956	.758	.918	.459	.613
Flour	.884	.945	.842	.907	.600	.674
Furniture	.929	.954	.855	.894	.613	.662
Mattress	.903	.941	.800	.888	.568	.646
Mineral	.780	.879	.752	.874	.497	.564
Mines	.812	.872	.802	.855	.504	.560
Oil	.811	.955	.876	.940	.555	.676
Paper	.637	.696	.528	.626	.349	.401
Printing	.659	.696	.638	.717	.481	.521
Textiles	.757	.773	.745	.794	.504	.542

tion to be used in the statistical analysis. The injury claims are broken down into categories for death (*DEAD*), dismemberment and disfiguration (*PD*), permanent partial (*PP*), and temporary total (*TT*). Two aspects of the injury data make them especially attractive to use in empirical analysis. First, the claims categories do not change over the sample period, and unlike most other states, the minimum and maximum weekly compensation levels as well as nominal percentage compensation are invariant with respect to the type of injury in any specified year. This greatly facilitates the empirical estimation of the effects of levels of compensation. Second, since the study is for a single state with relatively little secular change in the workers' compensation laws, much of the interstate variance in institutional aspects of the system that could contaminate a cross-sectional study has been eliminated.

I have employed the industrial data in an earlier study. They have been checked both for internal consistency and against census of manufacturing information. The data contain information on firm size, number of days in operation, and output figures that will be useful when removing the partial correlation between industrial accident rates and levels of benefits. Also, the wages and employment figures given are for production workers, those most affected by changes in the workers' compensation program, and moreover, are broken down by race and sex. So the data report on a fairly homogeneous group of nonunionized workers, working in quite disparate industries and experiencing quite varied extremes in industrial expansion. In addition, I have drawn on earlier research to calculate measures of human capital (accumulated educational expenditures per worker) for these workers (Butler 1979).

Before considering any of the empirical results, it will be useful to review how the workers' compensation system works in South Carolina. Consider the wage distribution for a typical industry in South Carolina, as shown in figure 3.1. The state has a minimum (*MINWC*) and maximum (*MAXWC*) weekly compensation and a nominal percentage compensation (*NPC*) ratio. The *NPC* ratio (NPC=60 percent) represents the proportion of wages that are replaced by benefits in the event of an injury, but of course is only effective between the minimum and maximum payments. Hence, all those with wages less than *MINWC/NPC* receive the minimum compensation, and those with wages greater than *MAXWC/NPC* receive the maximum payment. Given the structure of the South Carolina workers' compensation system with its relatively low maximum weekly compensation, the actual potential wage replacement is, in fact, much smaller than the 60 percent of wages that it nominally was during most of the sample period. The proportion of wages replaced for the representative worker in the industry illustrated in figure 3.1 would be $k \cdot 0.60$ where k is usually less than one. The value k is known as a factor limit, and is given in table 3.1 for the fifteen industries for 1950, 1960, and 1970. The factor limit is often used in policy evaluation when considering how adequate benefits are, but obviously also provides a picture of aggregate incentives to be safe (consider, for instance, the potential difference between a k of 0 and a k of 1.5). Although these calculations are often made for policy purposes, I am not aware of any research that has attempted to adjust k to account for the fact that workers' compensation benefits are tax-free. These calculations are contained in table 3.1. The no-tax data have *not* been

FIGURE 3.1
Wage Distribution
for a Typical Industry,
South Carolina

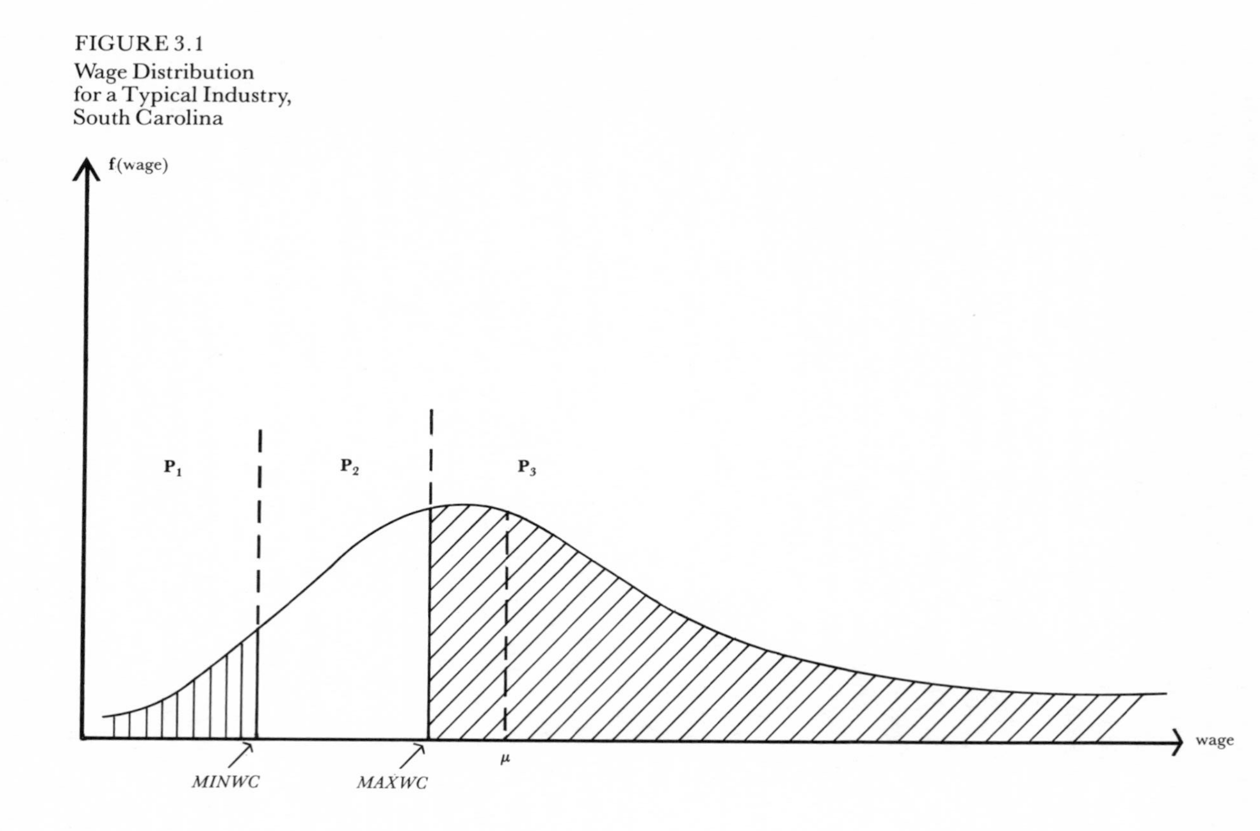

adjusted for federal, state, or social security payroll taxes. Since *k* in any given year is a nonlinear function of the wage distribution, the higher wage paper and printing industries will have a lower *k*, and the lower wage lumber and wood industries will have the highest *k*. What differentiates South Carolina from the rest of the United States is the declining value of *k* over time. As it declines, the tax adjustment becomes increasingly significant as the relative difference between the tax and no-tax columns increases (i.e., the tax/no tax factor limit ratio falls across all industries). This is why I use net wages (adjusted for federal, state, and social security taxes) in the analysis below.

Another view of the wage-benefit structure is presented in table 3.2. It gives average *after*-tax wages and expected weekly benefit payments (so that the ratio $WC/WAGE$ equals $k \cdot 60\%$). Again, this is a picture of secularly declining workers' compensation benefits. The mean values of various types of indemnity injury rates are also given in table 3.2, and together with the trends in benefits, provide support for the anecdotal wisdom that the workplace injury claims are responsive to changes in injury pay. Almost invariably, the claim frequencies fall as benefits fall and rise as benefits increase. What is especially intriguing is that the simple correlation seems strong even during the 1960s when injury rates were increasing sharply elsewhere in the United States. The strong positive correlation between claims and rate of indemnity pay is also consistent with the employee-dominant model, since if the firm's incentives were stronger, a negative (or at least a zero) correlation between claims and benefits would be observed. Since, however, other important secular changes were also taking place, causal relationships are extremely difficult to infer from these data alone. Hence, I use regression techniques to estimate models of wage and injury rate responses.

One important caveat concerning the data should be offered before presenting the empirical findings. The injury data reported are for workers' compensation cases which originated and closed during the respective fiscal year. The number and cost of cases originating in prior years and closed during the current fiscal year are given for most years, but are not broken down by industries. These data indicate that slightly less than half the cases were closed in the year in which they originated, and that this proportion remains fairly stable across time. While this implies that the actual benefits reflect full com-

TABLE 3.2
Trends in Benefits and Claims Frequencies, South Carolina

	Death		*PD*		PP				
	BEN	*IR*	*BEN*	*IR*	*BEN*	*IR*			
1940	$ 2,983	.682	$580	7.45	$ 301	3.65			
1945	9,710	.854	671	19.7	1,125	11.96			
1950	6,757	.825	544	13.1	712	5.88			
1955	7,086	.596	213	6.9	482	2.66			
1960	10,307	.033	264	5.7	1,096	4.40			
1965	9,484	.015	303	6.5	540	3.95			
1970	8,739	.290	211	5.5	627	4.24			
	TT		Indices						
	BEN	*IR*	*BEN*	*IR*	Wage	Workers' Comp.	Days Lost*		
1940	$176	173	$1,130	158	$36.57	$49.88	2,877		
1945	237	92	3,330	109	51.77	42.69	3,795		
1950	310	40	3,124	52	53.85	31.04	1,448		
1955	339	18	1,354	25	62.63	27.48	652		
1960	295	22	1,914	28	67.32	36.35	630		
1965	278	20	1,312	27	75.79	35.05	595		
1970	320	22	1,382	28	85.11	36.79	530		

Notes: The annual injury claims rate per 10,000 production employees includes only those industries in our sample. Benefits are average indemnity payments per claim. All benefits and wages are measured in 1967 dollars.

* Annual days lost due to injuries per 10,000 production employees.

pensation awards fairly adequately, the computed injury rates will be systematically understated (see table 3.2). I assume that this understatement, which results simply from the accounting and review process of South Carolina's Industrial Commission, is invariant over time (or random and uncorrelated with benefits). So instead of using the true injury rate π, my empirical work employs $\phi \cdot \pi$ (where ϕ is a constant). Under this assumption, neither the computed injury rate elasticities nor the compensating wages will be biased. Nonrandom

TABLE 3.3
Variables Used in
Injury and Wage Equations

Dependent Variables	
IRDAY	Number of days lost per employee due to injuries
IRDEAD	Annual injuries per employee resulting in death
IRPD	Annual injuries per employee resulting in permanent dismemberment or disfigurement
IRPP	Annual injuries per employee resulting in permanent partial injuries
IRTT	Annual injuries per employee resulting in temporary total injuries
PCIR	First principal component of IRDEAD, IRPD, IRPP, and IRTT
WG	After-tax weekly male wage of production employees
Independent Variables	
BLACK	Ratio of black to white male production employees
DAY	Average number of days the industry was operating
DVAL	Industry-specific measure of tightness of the product market
HKM	Quality-adjusted measure of human capital (real accumulated expenditures for the average worker in South Carolina, adjusted for migration and age distributions)
HKMSQ	Square of the HKM variable
IBDEAD	Average real annual payments for injuries resulting in death
IBPD	Average real annual payments for injuries resulting in permanent dismemberment or disfigurement
IBPP	Average real annual payments for injuries resulting in partial injuries
IBTT	Average real annual payments for injuries resulting in temporary total injuries
PCIB	First principal component of IBDEAD, IBPD, IBPP and IBTT
SIZE	Number of employees per establishment
TIME	Time trend variable
WC	Expected worker's compensation (for the average worker, see text)

variation on ϕ is likely to work against finding a significant benefit effect. For example, as benefits increase suppose a greater proportion of claims (say, more serious claims) *do not* close in the year in which they originate, then this will induce some negative correlation between the injury claim rates and benefits. In this case, the correlation of benefits and claim rates will appear smaller than it is and likely understate the effect of benefits on claim rates.

Analysis of Injury Rates

The empirical implementation of my claim rates model is given in table 3.4 separately for each class of compensable injury.[6] The theoretical expectation of the variables *DAY* (number of days per year the average plant was in operation as a measure of the pace of industrial activity) and *SIZE* (the number of employees per firm to capture economies of scale in the provision of safety equipment) are fairly well worked out (see Chelius 1974). As the pace of employee activity increases or the size of the firm decreases, accidents tend to increase as expected. Surprisingly, as the proportion of black males increases relative to white males, the injury rate falls. One would expect just the opposite if the *BLACK* variable is interpreted as a composition variable — blacks having the less skilled and more accident prone jobs. My best guess is that, since black employment has increased most in the older, established industries, the *BLACK* variable is really acting as a proxy for lack of industrial expansion.[7] The resulting sign on the human capital variable *HKM* was also puzzling, but may simply be capturing a younger work force with both more years of education and an education that was more expensive.

The variables of primary interest, wage (*WG,* the instrumented wage variable) and benefits (*IBDEAD, IBPD, IPBB,* and *IBTT* are respectively real benefits per case for death, permanent dismemberment or disfigurement, permanent partial and temporary total injuries), generally have the expected sign under the hypothesis of employee-dominant responses. The full set of benefit variables is included in each equation, since I have assumed that individuals are willing to accept greater risk of injury as benefits increase, and so the probability of any one kind of accident outcome will be associated

---

6. The model of individual behavior in equation 3.2 can, with the appropriate error assumption for ϵ_1 and ϵ_2, be written as a linear probability model. Summing both sides of the equation and dividing by n yields the injury rate regression. Note that this allows for a random coefficients interpretation of the model — the estimated coefficients yield the mean of the coefficient distribution if the coefficients are uncorrelated with the regressors.

7. Another explanation is that the negative sign represents an innovation effect — those firms hiring more blacks being also the more innovative in provision of modern safety equipment as well. However, a check of the crude employment ratios by industry does not bear this out. It is consistently the older, slower growing industries that employ relatively more blacks (see Butler 1979, table 18).

TABLE 3.4
Disaggregated
Injury Rate Regressions
(absolute t-statistic)

Independent Variables	(4a) *IRDEAD*	(4b) *IRPD*	(4c) *IRPP*	(4d) *IRTT*
Intercept	11.68	5.29	1.68	−67.95
	(2.39)	(0.30)	(0.02)	(0.42)
$\hat{WG}$	−.3785	−.4479	−.2449	−3.451
	(3.28)	(1.18)	(1.04)	(0.91)
IBDEAD	.0004	.0018	.0008	.0052
	(3.14)	(3.99)	(3.19)	(1.32)
IBPD	−.0013	−.0018	−.0017	−.0274
	(2.29)	(0.82)	(1.42)	(1.43)
IBPP	.0002	.0002	.0003	.0080
	(0.60)	(0.23)	(0.67)	(0.96)
IBTT	.0012	.0061	.0049	−.0482
	(0.98)	(1.32)	(1.90)	(1.15)
DAY	.0132	.0130	.0133	.5980
	(1.36)	(0.37)	(0.67)	(1.87)
BLACK	−.3674	−1.414	−.7721	−10.38
	(1.67)	(1.76)	(1.72)	(1.44)
HKM	.0058	.0223	.0090	.4076
	(2.72)	(2.87)	(2.07)	(5.81)
SIZE	−.0041	−.0026	.0026	−.0785
	(2.28)	(0.39)	(0.70)	(1.33)
TIME	.3061	−.325	−.0877	−11.18
	(2.18)	(0.63)	(0.30)	(2.42)
R^2	.135	.646	.665	.463
$F(3,440)$	3.58	2.29	2.54	1.58

Note: All coefficients have been multiplied by 10^3. Dummy variables for each industry were also included in the regressions, as were benefit dummies for observations when there were no injuries (and hence, no observed payments). The F-test statistics at the bottom represent tests of the hypothesis of no significant difference in the size of the estimated benefit effects in the respective equations.

with the full distribution of benefits. Throughout this investigation I take the level of benefits as given (as exogenous) and look at the simultaneous interaction of them with injury rates and wages. Hence, all the injury rate and wage equations are estimated by two-stage least squares (2SLS), the instruments being the exogenous variables in both equations.

At the bottom of table 3.4, values of the F-test statistic of equality of the benefit variables are recorded. Except for the *IRDEAD* equation, the null hypothesis of equal coefficients cannot be rejected at the usual 5 percent significance level. These coefficients are also found to be significantly different from zero at the usual levels. In this linear specification, this means that the elasticity of any type injury by benefits varies systematically from death to temporary total benefits. Death has the greatest elasticity,

$$\left(\frac{\partial\, IR}{\partial\, IBDEAD} \cdot \frac{IBDEAD}{IR}\right),$$

because the

$$\frac{IBDEAD}{IR}$$

ratio is highest. This is consistent with greater time intensity for the more severe injuries.[8] Not only do proportional changes in all benefits lead to differential injury rate responses by type of injury, but in all cases the implied elasticity is high for the more serious injuries. For example, the own elasticity (unrestricted) on the death benefits in equation 4a is 1.9. This tends to indicate that not all changes in the claim frequency are simply reporting changes. If only reporting changes were being recorded, I would not be observing the largest

---

8. As an example, consider a severe injury, (S), say, blindness, and a less severe injury, (NS), say, a broken arm, associated respectively with two states Z^S and Z^{NS}. Then assume that each has the same price elasticity

$$\frac{\partial lnZ^i}{\partial ln\pi^i} (i = S, NS).$$

The full price π^i has both a goods (with price P) and a time component (with price B, where B is the benefit level). Then

$$\frac{\partial lnZ^i}{\partial lnB^i} = \frac{\partial lnZ^i}{\partial ln\pi^i} \; \frac{\partial ln\pi^i}{\partial lnB^i} = \frac{\partial lnZ^i}{\partial ln\pi^i} \cdot \gamma_i,$$

$$\text{where } \gamma_i = \frac{B^i t}{\pi^i \cdot Z}$$

By the assumption above,

$$\frac{\partial lnZ^S}{\partial lnB^S} > \frac{\partial lnZ^{NS}}{\partial lnB^{NS}} \text{ since } \left(\frac{t}{X}\right)^S > \left(\frac{t}{X}\right)^{NS}$$

increases in claims as benefits rise in the more dangerous death and permanent partial categories. Additional support for the safety interpretation of claim changes will be found later in the compensating wage regressions. The significant wage differentials for risk of injury would not be observed if all changes in claims reflected only reporting propensities.

Given that the benefit coefficients are not significantly different (except in the *IRDEAD* equation), it is reasonable to seek a way of collapsing the injury benefit information into a single index to estimate the benefit effect more efficiently. Restricting the coefficients to be equal has that effect, but results with another type of index are reported in table 3.5. The problem is not only achieving a reasonable index for benefits but for the injury rates as well, since it will be necessary to capture the injury rate effects in a single index in the wage equation (simultaneity between *IR* and *WG* mandates a 2SLS procedure, but separate instruments are not available for each type of injury rate). Here, I choose to use principal components analysis to get the injury rate and injury benefit indices (respectively, *PCIR* and *PCIB*). The idea behind principal components is to find the linear combination that captures the greatest variance in the benefits (or injury) rates. This is, then, a statistical decomposition that forms the index number that captures the greatest variance in the benefit (or injury) variables. The resulting components had positive weights for all variables and captured 42 percent of the sample variation among the benefit variables and 64 percent of the variation among the injury rate variables.

Again, the signs of the nonbenefit regressors in table 3.5 show a fairly consistent pattern of coefficients that is congruent with theoretical expectations and the earlier empirical results of Smith (1974) and Chelius (1974). In the first four equations, the disaggregated injury rates are regressed on the benefit index. In the fifth equation the injury rate index is regressed on the benefit index and in the sixth, on the actual average benefits paid. The seventh and eighth equations provide results for another index of injury rates, the number of days lost per employee due to accidents. The results are similar to those in table 3.4.

The results using the benefit index tend to confirm the earlier results on the relative rankings of elasticity of the response in injury

TABLE 3.5
Injury Rate Regressions with Injury Rate and Benefit Indices (absolute t-statistic)

Independent Variables	(5a) *IRDEAD*	(5b) *IRPD*	(5c) *IRPP*	(5d) *IRTT*
Intercept	14.81	16.90	6.551	−58.52
	(2.91)	(0.91)	(0.63)	(0.36)
$\hat{WG}$	−.3923	−.5757	−.2730	−3.413
	(3.35)	(1.35)	(1.15)	(0.91)
PCIB	.0002	.0016	.0007	.0031
	(1.58)	(2.94)	(2.47)	(0.64)
IBDEAD	. . .	. . .	. . .	. . .
IBPD	. . .	. . .	. . .	. . .
IBPP	. . .	. . .	. . .	. . .
IBTT	. . .	. . .	. . .	. . .
DAY	.0097	.0072	.0058	.5775
	(1.03)	(0.21)	(0.31)	(1.90)
BLACK	−.3352	−1.345	−.7308	−9.575
	(1.51)	(1.67)	(1.62)	(1.34)
HKM	.0047	.0196	.0068	.3979
	(2.27)	(2.62)	(1.64)	(6.00)
SIZE	−.0040	−.0021	.0030	−.0844
	(2.18)	(0.31)	(0.83)	(1.43)
TIME	.3946	−.0437	.0821	−10.60
	(2.63)	(0.08)	(0.27)	(2.19)
R^2	.120	.634	.655	.457
$F(3,440)$				

rates. The implied elasticity (*dln* injury)/(*dln PCIB*), calculated at the sample means are 1.13, 0.36, 0.33 and 0.13 for 5a, 5b, 5c, and 5d, respectively. The death rate is most responsive to changes in the benefits and temporary total benefits are the least responsive.[9] Also note that the elasticities for 5e and 5g are 0.18 and 0.56 respectively. The

9. The lower elasticities for injuries not resulting in death may explain why Smith (1974) and others find insignificant wage premiums for nonfatal injuries.

TABLE 3.5 (continued)

Independent Variables	(5e) *PCIR*	(5f) *PCIR*	(5g) *IRDAY*	(5h) *IRDAY*
Intercept	−24.72	−50.46	−321.7	−525.6
	(0.16)	(0.33)	(0.82)	(1.37)
$\hat{WG}$	−3.780	−3.708	−9.196	1.758
	(1.06)	(1.04)	(0.10)	(0.02)
PCIB	.0049	...	.3533	...
	(1.08)		(3.08)	
IBDEAD	...	.0069	...	.3662
		(1.86)		(3.87)
IBPD	...	−.0268	...	−.628
		(1.49)		(1.37)
IBPP	...	.0074	...	.2045
		(0.94)		(1.03)
IBTT	...	−.0302	...	.7045
		(0.77)		(0.71)
DAY	.5050	.5355	5.691	7.341
	(1.76)	(1.78)	(0.78)	(0.96)
BLACK	−10.18	−10.98	−30.82	−51.65
	(1.51)	(1.62)	(0.18)	(0.30)
HKM	.3643	.3774	6.428	7.065
	(5.84)	(5.72)	(4.04)	(4.22)
SIZE	−.0715	−.0675	−.7926	−.8347
	(1.28)	(1.21)	(0.56)	(0.59)
TIME	−8.908	−9.843	−262.5	−317.4
	(1.95)	(2.26)	(2.26)	(2.88)
R^2	.524	.530	.461	.473
$F(3,440)$		1.577		2.06

Note: All coefficients have been multiplied by 10^3. Dummy variables for each industry were also included in the regressions, as were benefit dummies for observations when there were no injuries (and hence, no observed payments). The F-test statistics at the bottom represent tests of the hypothesis of no significant difference in the size of the estimated benefit effects in the respective equations.

elasticity for *IRDAY,* a variable reflecting both the frequency of accidents and the length of the recovery period, is larger than for nonfatal accidents. While the purpose of this analysis is to examine the effects a liberalization of workers' compensation benefits has on the frequency of accidents, we should note that higher injury pay may increase the length of recuperation and hence raise the cost of a workers' compensation system. Clearly, this is an important issue that future research should address.

TABLE 3.6
Injury Rate Regressions with Expected Benefits (absolute *t*-statistic)

Independent Variables	(6a) *PCIR*	(6b) *IRDAY*	(6c) *IRDEAD*
Intercept	−99.7	1089.0	.238
	(1.34)	(0.56)	(0.11)
WG	.015	.222	.006
	(1.48)	(0.87)	(2.04)
DAY	.388	6.51	−.003
	(1.44)	(0.93)	(0.43)
BLACK	−5.45	16.0	−.001
	(0.92)	(0.10)	(0.08)
HKM	.220	4.14	−.003
	(4.05)	(2.93)	(0.17)
TIME	−9.87	−2.07	−.0001
	(5.63)	(4.54)	(0.00)
SIZE	−.114	−1.94	−.002
	(2.68)	(1.76)	(1.25)
R^2	.521	.443	.129

Note: All coefficients have been multiplied by 10^3. Dummy variables for each industry were also included in the regressions, as were benefit dummies for observations when there were no injuries (and hence, no observed payments). The *F*-test statistics at the bottom represent tests of the hypothesis of no significant difference in the size of the estimated benefit effects in the respective equations.

Before discussing the implications of these results, an important caveat is in order. The use of actual benefits paid as the workers' compensation variable in tables 3.4 and 3.5 may be criticized as it is an *ex post* measure that may reflect workers of other than average characteristics (i.e., wages). The problem is that actual payments may confound the benefit parameters with the outcome of changes in those parameters. For example, raising the minimum benefit payment will disproportionately affect those with the lowest wages, and hence, there will be a dampening effect that may cause *actual* benefits to understate changes in *expected* benefits. An alternative measure would be the expected benefits that the "average" worker receives if injured. This variable, *WC*, was constructed by taking the average observed wages and then computing the expected benefits. Referring back to figure 3.1, *WC* is then simply the proportion of workers getting the minimum payment (*MINWC*) times *MINWC* ($P1 \cdot MINWC$)

TABLE 3.7
Estimated Claim Rate Elasticities with Respect to Benefits (equation number)

Benefit Measure	*IRDEAD*	*IRPD*	*IRPP*	*IRTT*	*PCIR*
PCIB	1.13 (5a)	.36 (5b)	.33 (5c)	.13 (5d)	.29 (5e)
WC	5.20 (6a)	...	...	...	1.02 (6c)

plus the proportion getting the maximum (*MAXWC*) times *MAXWC* (*P3·MAXWC*) added to the proportion in between these bounds times their average benefits ($P2 \cdot NPC \cdot \overline{W}$, p.113).[10] The results in table 3.6 employ this measure while excluding the wage regressor from the equation. This is because there was so little variation in the parameters of the benefit schedule (*MINWC, MAXWC,* or *NPC*) throughout the sample period that benefits are just a nonlinear function of the wage and under these conditions are difficult to statistically distinguish from the wage effect; there is a multi-collinearity problem. The reason for presenting a wide variety of specifications for the injury rate equation is to demonstrate the apparent robustness of the injury rate claim-benefit effect: claim frequency does appear to be responsive to changes in the benefit structure.

What do these results imply about the role the secularly diminishing indemnity payments play in the observed decline in claim rates? Focusing on the results employing my index measure of benefits (the equations with all benefits entered as separate regressors do not allow a concise interpretation), the respective elasticities are given in table 3.7. Here, the higher elasticities for the expected benefit variable *WC* in part captured the effect of the excluded wage variable (which, recall, is highly collinear with the *WC* variable in this sample). Using, for example, the 1.13 elasticity for *IRDEAD,* reveals that between 1950 and 1965 (these illustrative years were chosen in order

10. That is, I used the "standard" wage distribution and techniques outlined in Appendix A, "Limited Factor Calculations" (Berkowitz 1973a). I am very grateful to John Burton for making me aware of this reference and technique.

TABLE 3.8
Wage Equations
with Actual Benefits
(absolute *t*-statistic)

Independent Variables	(7a-1) *PCIR*	(7a-2) *DAY*	(7a-3) *DEAD*
Intercept	40.591 (12.58)	40.67 (9.61)	38.75 (6.97)
$PC\hat{I}R$	6.967 (0.24)	. . .	. . .
$PC\hat{I}RSQ$	115.15 (2.77)	. . .	. . .
$IR\hat{D}AY$	. . .	1.288 (0.75)	. . .
$IRD\hat{A}YSQ$	. . .	.2180 (2.13)	. . .
$IRD\hat{E}AD$	. . .	. . .	2101.8 (0.79)
$IRDE\hat{A}DSQ$	. . .	. . .	276428.7 (0.56)
IBDEAD	−.0001 (0.28)	−.0008 (1.16)	−.0008 (0.99)
IBPD	−.0012 (0.90)	.00004 (0.02)	−.0007 (0.31)
IBPP	.0001 (0.29)	−.0007 (0.93)	−.0003 (0.38)
IBTT	−.0007 (0.27)	−.0021 (0.71)	−.0030 (0.71)
$IBDEAD*\hat{Z}$	−.0004 (0.52)	−.00002 (0.59)	−.1047 (0.56)
$IBPD*\hat{Z}$	−.0125 (1.15)	−.0009 (1.60)	−.5188 (0.61)
$IBPP*\hat{Z}$	−.0009 (0.17)	.0001 (0.46)	.0912 (0.20)
$IBTT*\hat{Z}$	.0143 (0.56)	.0005 (0.35)	−.0558 (0.02)
$HKM*\hat{Z}$	.0633 (3.23)	.0025 (2.22)	3.862 (2.25)
$BLACK*\hat{Z}$	5.936 (2.00)	.2682 (1.78)	462.9 (2.14)
$DVAL*\hat{Z}$	19.32 (1.15)	.3374 (0.41)	1157.9 (0.56)
HKM	.0113 (1.43)	.0221 (2.54)	.0300 (2.62)
HKMSQ	−.00001 (2.55)	−.00002 (2.47)	−.00001 (2.06)

TABLE 3.8 (continued)

Independent Variables	(7a-1) *PCIR*	(7a-2) *DAY*	(7a-3) *DEAD*
BLACK	.2786	.1804	.4076
	(0.57)	(0.26)	(0.57)
DVAL	6.262	4.394	18.46
	(1.58)	(2.14)	(3.06)
TIME	1.674	1.775	1.137
	(0.32)	(4.19)	(4.89)
R^2	.941	.915	.870

Note: Industrial and benefit dummy variables were also included in the regression model (see notes to table 3.4 although the coefficients have not been rescaled). $\hat{Z}$ stands for *PCIR, IRDAY,* and *IRDEAD* in their respective equations.

to avoid either the years of World War II or the expansionist period of the late 1960s) the death claim rate fell by about 400 percent, and the benefit index fell by about 82 percent. This implies that about 23 percent ((0.82·1.13)/4.00=0.23) of the decrease is explainable in terms of benefit changes. Similar calculations based on the *PC/B* elasticities yield the fraction of the decrease explained by benefits for *IRPD, IRPP, IRTT* and *PC/R* respectively as 50 percent, 97 percent, 97 percent, and 36 percent. While these results must be regarded as tentative, given the perhaps unique time period and the state they represent, they are suggestive enough to warrant further research. What they seem to indicate clearly is that the recent increase in claim rate frequency is in part — in fact, substantial part — the result of more liberalized workers' compensation benefits. Moreover, it appears that all the change is not in the reporting of injuries, but in the change in the safety behavior that the injury pay system may induce in workers.

Wage Responses to Injuries and Post-Injury Payments

In this section the impact of post-injury benefits, injury rates, human capital, and industrial expansion on after-tax wages is examined. Again, I present alternative estimates to emphasize the robustness of the findings. Tables 3.8 to 3.10 present regressions using the average production wages for males on several industrial and worker charac-

TABLE 3.9
Wage Equations
with the Benefit Index
(absolute *t*-statistic)

Independent Variables	(7b-1)	(7b-2)	(7b-3)
Intercept	41.60	40.35	36.22
	(13.84)	(11.53)	(8.43)
$PC\hat{I}R$	1.011	...	...
	(0.04)		
$PC\hat{I}RSQ$	109.40	...	...
	(2.94)		
$IR\hat{D}AY$	...	1.120	...
		(0.60)	
$IRD\hat{A}YSQ$	...	.2039	...
		(2.27)	
$IRD\hat{E}AD$	...	...	1729.8
			(0.65)
$IRDE\hat{A}DSQ$	...	...	367385.3
			(1.08)
PCIB	−.0003	−.0012	−.001
	(1.15)	(1.71)	(1.50)
$HKM*\hat{Z}$	.0661	.0027	4.472
	(3.51)	(2.66)	(3.19)
$PCIB*\hat{Z}$	−.0017	−.0001	−.2323
	(1.26)	(1.32)	(1.08)
$BLACK*\hat{Z}$	7.437	.3480	529.7
	(2.58)	(2.49)	(2.80)
$DVAL*\hat{Z}$	22.27	.4767	1479.3
	(1.38)	(0.65)	(0.86)
HKM	.0093	.0206	.0305
	(1.19)	(2.46)	(2.88)
HKMSQ	−.00001	−.00002	−.00001
	(2.48)	(2.32)	(2.22)
BLACK	.1577	.0122	.3983
	(0.29)	(0.02)	(0.62)
DVAL	6.016	4.297	18.95
	(3.85)	(2.22)	(3.03)
TIME	1.719	1.806	1.110
	(5.56)	(4.60)	(7.4)
R^2	.941	.921	.891

Note: Industrial and benefit dummy variables were also included in the regression model (see notes to table 3.4 although the coefficients have not been rescaled). $\hat{Z}$ stands for *PCIR, IRDAY,* and *IRDEAD* in th ir respective equations.

teristics.[11] The time trend variable plays its usual ubiquitous roles (picking up secular changes in industrial and safety technology, for instance), and *DVAL* is a labor market tightness variable being measured as the deviations of gross sales (of each industry) from its trend line. *BLACK* picks up the composition of blacks in the labor force. The effects of formal schooling are captured by the *HKM* variable (see Butler 1979 for a discussion of this variable). No on-the-job training variable is included, but the age distribution for these production workers across the 1940, 1950, 1960 and 1970 censuses is sufficiently stable that an aggregate training variable would show very little variance at any rate. The variables of interest are the benefits and the probability of an accident. One expects that as post-injury benefits rise, wages fall, *ceteris paribus.* No matter how the benefits are modeled, this appears to be the case (whether one uses the full distribution, table 3.8, the benefit "index," table 3.9, or expected benefits, table 3.10).[12]

Because of the simultaneity of wages and injury rates and the hedonic effect that an increasing probability of death or injury has on wages, the injury claim variable has again been instrumented with all the exogenous variables in both the wage and injury rate equations, and this predicted injury rate variable has been included, along with interactions of this variable with the other included exogenous variables (the hedonic interpretation suggesting nonlinearities between the wage and injury rate), in the wage equation. This is necessitated in a properly specified equation because the equilibrium wage-injury rate locus is swept by the sorting of individuals and firms on the basis of the exogenous shift variables. Excluding the interaction of these shift variables with the hedonic characteristics (injury rates in this example) induces a sorting or selectivity bias into the estimates.[13] The specifications in tables 3.8, 3.9, and 3.10 give the wage equations estimated

---

11. The dependent variable is actually the *after-tax* weekly wage, being adjusted for state, federal and social security taxes. Estimates without the tax adjustment were virtualy identical to those reported here.

12. Since the probability of injury is less than one, we would expect that a dollar increase in benefits results in less than a dollar decrease in wages. This is the case. In equations 7c-1, 7c-2, and 7c-3, wages decrease respectively by 11.5, 14.0 and 13.7 cents with each dollar increase in expected benefits.

13. See Butler (1981) for the reasoning behind this approach to estimating risk premiums.

TABLE 3.10
Wage Equations
with the Expected Benefits

Independent Variables	(7c-1)	(7c-2)	(7c-3)
Intercept	40.86	36.05	37.15
	(9.40)	(5.22)	(6.18)
$\hat{PCIR}$	52.45	...	...
	(1.12)		
$\hat{PCIRSQ}$	92.61	...	...
	(1.71)		
$\hat{IRDAY}$	...	3.441	...
		(1.03)	
$\hat{IRDAYSQ}$	...	.1185	...
		(0.83)	
$\hat{IRDEAD}$	...	...	3108.1
			(0.98)
$\hat{IRDEADSQ}$	...	...	120745.2
			(0.32)
WC	−.0018	−.0024	−.0025
	(1.69)	(1.47)	(1.31)
$WC*\hat{Z}$	−.0088	−.0003	−.5721
	(1.19)	(0.82)	(1.03)
$HKM*\hat{Z}$	.0516	.0016	3.142
	(1.91)	(1.00)	(1.79)
$BLACK*\hat{Z}$	4.670	.2005	401.98
	(1.16)	(0.88)	(1.54)
$DVAL*\hat{Z}$	3.553	−.1811	443.55
	(0.16)	(0.14)	(0.20)
HKM	.0267	.0414	.0469
	(2.15)	(2.23)	(2.14)
HKMSQ	−.00002	−.00002	−.00001
	(2.28)	(1.59)	(1.64)
BLACK	.7467	.6308	.4559
	(0.86)	(0.50)	(0.45)
DVAL	6.551	4.197	14.17
	(3.00)	(1.15)	(1.86)
TIME	1.430	1.415	.6625
	(3.49)	(2.32)	(2.22)
R^2	.884	.788	.795

Note: Industrial and benefit dummy variables were also included in the regression model (see notes to table 3.4 although the coefficients have not been rescaled). $\hat{Z}$ stands for *PCIR, IRDAY,* and *IRDEAD* in their respective equations.

TABLE 3.11
Implied "Compensating" Wages

Benefit measure	Injury Rate Measure		
	PCIR	*IRDAY*	*IRDEAD*
Actual Benefits (*IBDEAD, IBPD, IBPP, IBTT*)	\$3,298 (7a-1)	\$182 (7a-2)	\$254,320 (7a-3)
Benefit Index (*PCIB*)	\$3,116 (7b-1)	\$183 (7b-2)	\$258,118 (7b-3)
Expected Benefits (*WC*)	\$3,407 (7c-1)	\$190 (7c-2)	\$178,854 (7c-3)

Note: Computed at the sample means (see text).

respectively with actual benefits, the benefit index, and expected benefits as the benefit effect. Each of these measures of injury payment is interacted with alternative specifications of the injury rate effect: the injury index (*PCIR*), the number of days lost due to accidents (*IRDAY*), and the probability of dying (*IRDEAD*). In all cases, wages increase at an increasing rate as the probability of industrial injury risk rises; the workers in this sample are apparently risk averse. Again, the existence of such premiums and changes in claim frequency indicates that they are not solely reflecting "reporting" changes.

The analysis found in tables 3.8, 3.9, and 3.10 is translated into the "compensating" wages found in table 3.11 (the numbers below each compensation figure refer to the wage equation upon which the calculations were made). For each incremental increase in the injury rate measure, wages increase by the indicated amount. For example, the analysis based on the actual distribution of benefits (*IBDEAD, IBPD, IBPP,* and *IBTT*) indicates that the firm needs to implicitly compensate an employee an extra \$182 for an added day lost due to injury (middle column in the first row of table 3.11). As the \$182 figure far exceeds the average daily wage, there appears to be a fairly sizable premium for the nonpecuniary aspects of injury risk. This is confirmed in the risk of death equations. Here, the market appears to be valuing a life — by the "willingness to pay" criterion (see Smith 1979a) — at \$180,000 to \$260,000. These figures are very close to the numbers Thaler and Rosen (1975) obtained using a quite different microsample of higher risk workers.[14] Not only are the workers in South Carolina responsive to benefit changes, they appear to be sensitive to changes in job riskiness as well.

Conclusion

This study is only an initial attempt to fill the void of empirical research concerning one of the most far-reaching and topical of social insurance programs, workers' compensation. Evidence from a data set uniquely suited to analyze workers' compensation tends to corroborate the model of employee behavior outlined — accident rates do appear responsive (in the hypothesized directions) to changes in the structure of benefits as well as wage shifts, and wages adjust both to changes in the probability of injury and to changes in post-accident benefits. That the frequency of accidents is affected by injury payments to employees partially explains why the cost of workers' compensation has risen in recent years. In future work, more recent and nationally representative samples should be used to check the universality of the results here. Should subsequent analyses bear out my finding of a significant claim frequency elasticity with respect to benefit changes, then policy analysts will need to assimilate these elasticities into their cost projections.

14. This interpretation is not strictly correct, as explained in Butler 1981.

4 · EMPLOYMENT HAZARDS AND FRINGE BENEFITS: FURTHER TESTS FOR COMPENSATING DIFFERENTIALS

Stuart Dorsey

The rising industrial injury rates in the 1960s fostered a widely held view that increasing workplace safety and compensating injured workers required government intervention. In the early 1970s the Occupational Safety and Health Act and the recommendations of the National Commission on State Workmen's Compensation Laws resulted from this view. The act mandated certain safety standards, while the recommendations led many states to substantially liberalize workers' compensation benefits. These efforts were justified by the perception that private, unregulated markets were unable to provide satisfactory solutions to the problems of hazardous work environments.

Economic theory predicts that, under certain conditions, the market will produce the optimal solution to the problem of employment hazards. If firms and workers have adequate information about job risks and transactions costs are low, firms have an incentive to supply the optimal level of safety, such that the value of further reduc-

This chapter was written before Mr. Dorsey assumed his position with the U.S. Senate's Committee on Finance and does not represent an official position of the committee. —J.D.W.

tions in risk is less than the real resource costs of achieving them. Dangerous occupations still will exist, but they will pay a premium wage and attract workers who prefer the higher wage-risk combination. Workers who are more risk-averse will choose safer employment at lower wages. By implication, minimum safety standards reduce the welfare of the former group, forcing them to buy more safety than desired.

At that time there was very little empirical evidence with which to judge the market's performance. Economists since have moved to correct this deficiency. Viscusi (1979a) found evidence supporting a number of hypotheses predicted by competitive theory. But the prediction of compensating differentials for increased risk of injury has received the most attention. Smith (1979a) provided a fine review of these studies, and Brown (1980), Olson (1981), and Dorsey and Walzer (1983) have made recent contributions.

This paper extends tests for compensating differentials by considering the extent to which risk premiums may appear in nonwage compensation. Theory requires that relatively high physical risk be accompanied by higher total compensation, not necessarily just wage premiums. This was made clear by Adam Smith, who wrote that "the whole of advantages and disadvantages of the different employments of labor and stock must, in the same neighborhood, be either perfectly equal or continuously tending toward equality." (1776, p. 151). Thus a compensating differential in *some* employment condition is required to maintain the attractiveness of hazardous jobs. But that premium need not take the form of a wage bonus. Indeed, it seems that workers in risky jobs would be likely to prefer fringe benefits that come into play in the event of injury or death, including pensions with disability provisions, liberal sick leave policies, and health, life, and accident insurance.[1] If there is a systematic, positive relationship between employment hazards and the nonwage-wage compensation ratio, wage premiums will understate the compensating differential. Since physical risk is itself a nonwage employment condition, it is logical to try to include as many other aspects of nonwage compensation as possible when testing for equalizing differences — especially

1. It is easily shown that if the risk of injury is small, risk-averse workers will prefer to receive a portion of the compensating differential in the form of insurance.

when about 15 percent of total employee compensation is in nonwage form (Chamber of Commerce 1977).

Testing this hypothesis is important for three reasons. First, the estimates provide a key test of how well unregulated labor markets handle the problem of employment hazards. The compensating differential does double duty: it provides the incentive for firms to reduce employment hazards, and it compensates employees for bearing extra risk. A second use of the estimates is in calculating the valuation of life. The shadow price of an increment in the risk of death (that is, the additional wage required to entice the marginal worker to accept this extra risk) is observed, and by extrapolation we can compute the amount a representative worker would pay to save a life.[2] Finally, these studies have made an important contribution to the body of evidence concerning the efficiency and optimality of unregulated labor markets in general. Each use is a reason for improving the estimation by extending the analysis to include nonwage compensation.

The present analysis will make estimates of the total compensating differential for employment hazards, including nonwage premiums. In addition, this study will test for trade-offs between workers' compensation costs and wage and nonwage compensation. The results of these tests will bear upon the question of who pays the cost of workers' compensation protection. All the empirical tests will focus on the costs of employee compensation to firms, unlike previous efforts that mainly have used the worker as the unit of observation.

Estimation

Previous studies focused on wage premiums (Allen 1981 is a notable exception). This was not due to oversight, but rather because the large microdata sets, which have proved so useful because of rich detail on personal characteristics, have not provided information on fringe benefits. In this study a cross section of establishment expenditures for employee compensation was used to estimate compensating

2. Care must be taken here because the observed wage-risk relationship reflects the wage premium required to attract the marginal worker. It does not represent the wage that a representative worker would require as compensation for the added risk. That is, the compensating differential does not trace out an indifference curve (see Smith 1979a).

differentials. The advantages of establishment data are that they allow a test of the theory from the perspective of firms' costs and permit a measure of total employee compensation, based upon expenditures for wages and fringe benefits.

The Employers' Expenditures for Employee Compensation (EEEC) survey for 1977 is the basic data file for this analysis. It reports total yearly expenditures for wages and employee benefits by component for more than three thousand nonfarm establishments.[3] The number of employees and the presence or absence of a collective bargaining agreement are also reported in the file. The data are listed separately for office and production workers. The EEEC survey was chosen because it allows computation of hourly fringe benefit costs and thus a measure of the "full wage." The major disadvantage of this file, however, is the absence of direct measures of employee characteristics that economists have found useful in empirical studies of wage determination. It is necessary to control for the age and education of the work force, as well as race and sex composition, in making comparisons across firms.

There is no entirely satisfactory solution to this problem, but following Freeman (1981) and Antos (1983), I merged the EEEC file with average worker characteristics by industry, using data from the May 1978 and May 1979 Current Population Survey (CPS). Mean education and age and the percentage nonwhite and female were computed by three-digit industry and assigned to firms accordingly. An additional control for work force characteristics is the ratio of production workers to total establishment employment. Firms with higher skilled production workers tend to have a greater percentage of office workers, so that a higher ratio would indicate a less skilled work force. These are crude controls. There may, for example, be important interindustry variations in employee characteristics. To my knowledge, however, this is the best available procedure, and the empirical results suggest that it does capture a significant portion of compensation variation across firms.

The job risk measures are nonfatal occupational injury rates and average number of lost workdays per incident. These provide a

3. The EEEC survey was conducted biennially from 1968 to 1976 and in 1977, the final year.

measure of the probability and severity of nonfatal injury. Also included in the analysis is the probability of fatal injury. The Bureau of Labor Statistics reports all these figures, and I have assigned them by three-digit SIC codes.[4] The well-known weakness of these measures is that they are not occupation-specific, resulting in a measurement error that may bias estimates. My previous work suggests that the wage premium tends to be understated as a result. But in any event, industry-specific variables are relevant here because the unit of observation is the firm. Of course, estimates may still be biased because firms will employ workers at different jobs and different risks. Assigning average injury data for all employees is likely to be particularly inaccurate for office workers. Some of this problem is eliminated by limiting the sample to firm expenditures for production workers.

A final sample of 1,885 observations on firm expenditures for production workers is used. The loss of observations is the result of inability to match injury rates and work force characteristic variables to all firms.

All variable definitions and means are reported in table 4.1. There are three dependent variables of interest: the straight-time average hourly wage, hourly expenditures for voluntary nonwage benefits, and total employee compensation per hour. These were computed by dividing total expenditures for each by total hours worked in the firm. Table 4.2 reports the coefficients obtained by regressing the natural logarithm of each against the explanatory variables. The wage equation turns out to be fairly well behaved. Each of the work force characteristic estimates has the expected sign and is statistically significant. *BCRATIO* has a negative coefficient, as expected, and collective bargaining exerts a strong, positive influence on wages — indeed, a bit too strong, given estimates from individual worker data. Part of the union coefficient probably represents the correlation of unionization and worker attributes (see Kalachek and Raines 1980). Given that these attributes are industry-specific, the firm-specific union variable would be a proxy for unmeasured variation in work

4. The SIC injury rates can be assigned directly because firms in the EEEC are classified by SIC. Previous studies (Olson 1981; Dorsey and Walzer 1983) use individual data, with workers classified by census codes. The matching of SIC and census coding resulted in a substantial loss of observations.

TABLE 4.1
Definitions of Variables

		Mean
Compensation Measures		
WAGE	Average hourly wage[1]	5.81
NWAGE	Hourly expenditures for nonwage compensation[2]	.92
FWAGE	WAGE + NWAGE	6.73
WCRATE	Workers' compensation costs per $100 of payroll	1.85
UI	State unemployment insurance costs per $100 of payroll	1.50
Work Force Characteristics by Three-Digit Industry		
ED	Mean years of education	13.2
AGE	Mean age	35.9
FEMALE	Percent female	53.7
NONWHITE	Percent nonwhite	10.3
Firm Characteristics		
UNION	Equals 1, if majority of work force is covered by collective bargaining agreement	.382
BCRATIO	Ratio of production workers to total establishment employment	.774
EMPLYS	Number of employees in establishment	485.1
20–49	Equals 1, if establishment employment is 20–49	.147
50–99	Equals 1, if establishment employment is 50–99	.117
100–499	Equals 1, if establishment employment is 100–499	.259
500–999	Equals 1, if establishment employment is 500–999	.098
1000+	Equals 1, if establishment employment is 1000 or more	.183
SMSA	Equals 1, if firm is located in a SMSA	.715
CORP	Equals 1, if establishment is part of a larger company or corporate enterprise	.406
Employment Hazard Measures by Three-Digit Industry		
FREQ	Probability of a nonfatal, lost-workday accident	3.63
SEV	Average number of lost workdays per lost-workday accident	16.53
FATAL	Probability $\times 10^3$ of a fatal work injury	.071

Notes: Also included in the analysis are four regional dummy variables.

1. Excludes premium pay for overtime hours and shift differentials.

2. Total expenditures per man-hour worked for pension funds, insurance, sick leave, overtime and shift differentials, vacations and holidays, and other voluntary fringes.

TABLE 4.2
Regression Estimates of Hourly Compensation
(absolute *t*-statistic)

Dependent Variable	ln(*WAGE*)	ln(*NWAGE*)*	ln(*FWAGE*)
Work Force Characteristics			
ED	.059	.152	.066
	(6.97)	(6.68)	(7.27)
AGE	.020	.075	.023
	(7.75)	(10.63)	(8.54)
FEMALE	−.002	−.0003	−.002
	(5.96)	(.28)	(5.03)
NONWHITE	−.017	−.035	−.019
	(12.01)	(9.04)	(12.29)
Firm Characteristics			
UNION	.178	.542	.220
	(11.69)	(13.39)	(8.54)
BCRATIO	−.108	−.721	−.164
	(3.33)	(8.24)	(4.70)
20–49	.100	.304	.126
	(4.71)	(5.01)	(5.52)
50–99	.096	.279	.128
	(4.10)	(4.28)	(5.10)
100–499	.164	.552	.207
	(7.84)	(9.37)	(9.20)
500–999	.180	.631	.236
	(6.68)	(8.52)	(8.14)
1,000+	.312	.847	.384
	(12.95)	(12.72)	(14.78)
SMSA	.064	.028	.066
	(4.54)	(.07)	(4.40)
CORP	.032	.174	.048
	(2.31)	(4.67)	(3.22)
Employment Hazard Measures			
FREQ	.031	.072	.034
	(9.45)	(8.13)	(9.57)
SEV	.017	.036	.019
	(6.44)	(5.05)	(6.68)
FATAL	.063	−.264	.050
	(.82)	(1.19)	(.60)
Constant	−.152	−6.02	−.319
Adj. R^2	.583	.575	.621
F-value	133.0	121.1	155.6
N	1885	1773	1885

Notes: Each regression included four regional dummies.

*Includes only firms with positive contributions to employee fringe benefits.

force quality. Establishment-size dummies show a premium paid by larger firms, consistent with recent analyses of individual wage differences (see Mellow 1982).

Of primary interest are the positive and significant coefficients for the frequency and severity of nonfatal injury. They imply that (1) a wage premium of 3 percent accompanies a 1 percent increase in the probability of injury and (2) an increase in the severity of injury of 1 lost workday per accident is associated with a 1.7 percent increase in the wage.

These estimates are fairly close to those obtained using individual employee data from the 1978 Current Population Survey (2.3 percent and 0.9 percent, respectively, in Dorsey and Walzer 1983). But I find no evidence of an earnings premium for elevated risk of death. The fatal injury premium is, as one would expect, highly correlated with the nonfatal risk measures, and therefore is quite sensitive to the presence of *FREQ* and *SEV*. When the latter are omitted, fatal risk does enter with a positive and significant coefficient.

These estimates may be biased indicators of the total compensating differential if the risk variables have markedly different effects on either the probability that a firm will offer fringe benefits or the generosity of employer contributions to such plans. A binary (0,1) variable, indicating whether or not the firm provided any type of nonwage benefit, was estimated as a function of the explanatory variables by the logit technique. Little of interest was found in the probability model, as few firms (6 percent) had no fringe benefit plans (these estimates are not shown). *FREQ* and *SEV* have very small but statistically significant, positive coefficients. More interesting is the second equation in table 4.2, in which the natural logarithm of hourly nonwage compensation is the dependent variable. Only those firms with positive expenditures for fringes are included in this sample. There we find a strong correlation between the nonfatal risk measures and nonwage compensation, substantially stronger than the influence that these variables exert on wages. Both the *FREQ* and *SEV* coefficients are more than doubled in the fringe benefit equation. The implications for the total compensating differential are apparent in the full wage equation. With total hourly expenditures for employee compensation as the dependent variable, the estimates on *FREQ* and *SEV* do increase, but modestly. Fatal risk remains statistically insignificant. Thus, while the nonfatal risk measures exert a stronger impact on nonwage compensation, wage premiums are not wildly biased estimates of the full-wage percentage compensating differential. They would, however, substantially understate the *absolute* amount of the premium received by workers in hazardous jobs, as the second equation indicates a significant compensating differential appearing in nonwage compensation.

The last two regressions have some other results of interest. Collective bargaining has a strong impact on nonwage compensation, confirming Freeman (1981). Consequently, the wage coefficient considerably understates the total compensation premium enjoyed by unionized employees. The coefficient on *BCRATIO* also rises markedly (in absolute value) in the full wage equation, suggesting that more highly skilled workers have a preference for nonwage benefits that could lead to downward bias in estimates of personal earnings differentials when wages alone are considered. Finally, older workers appear to have a preference for nonwage compensation; this is consistent with other evidence (Dorsey 1982 and Freeman 1981).

Other Specifications

Table 4.3 reports the results of additional tests of the compensating differential hypothesis. Equation 1 specified the relationship as a quadratic. No consistent pattern is found under this formulation: the estimates indicate an increasing premium for severity of injury, but a decreasing differential for frequency. The performance of the fatal risk variable is not improved. In equation 2, risk is interacted with collective bargaining. Many researchers (Olson 1981; Viscusi 1979a) have found job-risk premiums higher for union workers. But here we find conflicting evidence. The compensating differential for probability of nonfatal injury is apparently smaller in the union sector. At the same time, the insignificant effect of fatal risk in the full sample appears to be the result of two opposing tendencies: a negative differential in the nonunion sector and a positive premium for union workers. I find a larger premium for union workers only for elevated risk of death. These are qualitatively similar to the results found in Dorsey and Walzer (1983).

TABLE 4.3
Risk Coefficients
from Alternative Specifications
(dependent variable = ln(*FWAGE*)
(absolute *t*-statistic)

	(1)	(2)	(3)
FREQ	.081 (8.52)	.054 (11.75)	.042 (10.55)
SEV	.001 (.13)	.014 (3.41)	.017 (5.20)
FATAL	.138 (.81)	−.475 (4.12)	−.074 (.81)
$FREQ^2$	−.004 (4.57)	...	...
SEV^2	.0004 (1.72)	...	...
$FATAL^2$	−.060 (.38)	...	...
FREQ × *UNION*	...	−.024 (3.69)	...
SEV × *UNION*	...	.007 (1.32)	...
FATAL × *UNION*	...	.726 (4.39)	...
FREQ × *EMPLYS*	...	...	.42-7E (.02)
SEV × *EMPLYS*	...	...	.31-5E (1.75)
FATAL × *EMPLYS*	...	...	−.64-4E (.98)
Adj. R^2	.598	.600	.594

Note: All regressions included the independent variables listed in table 4.1.

The coefficients imply that an increase in the risk of death of 1 chance in 10,000 yields a 2.51 percent increase in total earnings for union workers. This is smaller than the corresponding premiums estimated by Olson and Dorsey and Walzer of 9.1 percent and 5.7 percent, respectively.

Finally, table 4.3 reveals that risk premiums are not related systematically to establishment size.

TABLE 4.4
Effect of Job Risk
on Specific Nonwage Benefits
(absolute *t*-statistic)

	Pensions		Holidays, Vacations		Sick Leave		Insurance	
	Provision (0,1)*	Expen/ hour	Provision (0,1)*	Expen/ hour	Provision (0,1)*	Expen/ hour	Provision (0,1)*	Expen/ hour
FREQ	.016	.027	.009	.057	−.006	.036	.0006	.062
	(3.42)	(2.08)	(2.88)	(7.22)	(1.28)	(2.00)	(4.23)	(6.67)
SEV	.013	.028	−.003	.019	−.012	.043	−.0002	.033
	(3.45)	(2.58)	(1.10)	(2.75)	(2.92)	(2.58)	(1.68)	(4.38)
FATAL	−.023	.087	−.026	−.047	...	...	−.0006	−.026
	(2.10)	(2.57)	(3.70)	2.34			(2.22)	(1.13)
R^2	...	.329	...	.538	...	.789	...	.372
(likelihood ratio)	(.420)	...	(.201)	...	(.211)	...	(.301)	...
N	1885	1198	1885	1733	1885	1080	1885	1629

Note: *Coefficients were obtained by multiple logit estimation and are the derivatives of the independent variables with respect to the dependent variable, evaluated at the formers' means.

Compensating Differentials by Type

This data set allows a more thorough exploration of how compensating differentials are formed than previously has been possible. Specifically, I estimated the effect of the risk measures on the probability of a firm providing a given benefit and expenditures for each. Coverage equations, with binary dependent variables for each fringe benefit, were estimated by logit. Table 4.4 lists the derivatives of the independent variables, evaluated at the independent variable means. Thus they indicate the marginal effect of the risk measure on the probability of the firm supplying this type of benefit. Next, the generosity of firms' expenditures for a specific fringe benefit was estimated, given that the firm did supply that type of benefit. Hourly expenditures for nonwage benefits were estimated by ordinary least squares. All other control variables shown in table 4.2 were included in each of the models.

A 1 unit increase in frequency of injury implies a 1.6 percent rise in the probability of the firm providing a pension, and given that it has, a 2.7 percent increase in contributions per hour. The injury severity estimates also are positive and significant. These results are expected because the disability provisions of many pensions should make an attractive compensation component in hazardous occupations. Turning to the other models, *FREQ* and *SEV* have positive effects on expenditures for each of the other fringe benefits. They are not, however, consistently related to the probability that any given benefit might be included in the compensation package. The effect of fatal risk is generally negative, with the only exception being expenditures for pensions.

Workers' Compensation Trade-offs

An important advantage of the EEEC data file is the availability of actual workers' compensation costs. Workers' compensation is a crucial theoretical factor in the determination of compensating differentials for risk of on-the-job injury, especially for nonfatal risk. The program essentially defines the liability for costs arising from workplace injuries. Existence of compensating differentials presupposes that the employee is liable for some of the costs of injury. But to the extent that liability is shifted to the employer, the required premium shrinks. Thus, holding physical risk constant, variations in the costs of workers' compensation will for the most part reflect differences in the generosity of benefits and should result in offsetting changes in wages or benefits.[5]

Using the 1978 Current Population Survey, Dorsey and Walzer (1983) estimated a dollar-for-dollar trade-off between wages and workers' compensation costs. The EEEC survey allows another test of this hypothesis from the perspective of the firm. An advantage here is that I was able to compute the actual workers' compensation

5. A caveat on the trade-off between workers' compensation costs and wages: to the extent that variations in cost reflect factors other than differences in benefits, the trade-off will be less than dollar-for-dollar. For example, if higher costs result from less efficient administration of the program, the wage firms would be willing to pay for a given number of workers would fall, but the wage required to attract workers into the job would not decline, there being no diminution in risk. Similarly, manual rating could raise the workers' compensation costs of a firm, even though its own safety record might be unchanged. This rise in costs, due to increased riskiness of other firms in the same class, would not be fully offset by reduced wages.

TABLE 4.5
Compensating Differential Estimates with Workers' Compensation Variable (absolute *t*-statistic)

Dependent Variable	ln(*WAGE*)	ln(*NWAGE*)	ln(*FWAGE*)	ln(*FWAGE*)*
FREQ	.034	.074	.037	.032
	(10.16)	(8.22)	(10.27)	(9.47)
SEV	.017	.037	.019	.017
	(6.57)	(5.11)	(6.81)	(6.20)
FATAL	.083	−.264	.071	.068
	(1.09)	(1.19)	(.86)	(.86)
WCRATE	−.0167	−.0148	−.0178	−.0098
	(4.17)	(1.30)	(4.14)	(2.35)
Adj. R^2	.583	.576	.625	.659
N	1885	1773	1885	1885

Note: *Includes a control for state unemployment insurance costs.

costs, rather than having to rely on manual insurance rates, which must be matched to individuals according to their occupation. Table 4.5 reports the estimates of risk and workers' compensation rate coefficients obtained from regressions with the dependent variables used in the original tests: the wage, hourly nonwage compensation, and total hourly compensation. The original set of controls again were included in these equations. The estimates clearly indicate a trade-off between compensation and workers' compensation costs, holding physical risk constant. The coefficient on *WCRATE* implies that a \$1.00 increase in workers' compensation costs per \$100.00 of payroll is associated with a 1.67 percent reduction in the hourly wage, or \$1.67 per \$100.00 of payroll. This estimate, however, is not significantly different from −0.01 at the 5 percent confidence level. These results are consistent with a complete trade-off and are remarkably close to those obtained with data on individuals. In the earlier paper (Dorsey and Walzer 1983), we estimated the coefficient of *WCRATE* to be −0.0142 for nonunion production workers. For purposes of comparison, the *WCRATE* estimate is −0.0121 (*t*-statistic=−2.26) when the present sample is limited to nonunion firms. Unlike the previous study, however, I do find a trade-off in the union sector as well, where the coefficient is −0.020 (*t*-statistic=−3.30).

There is not strong evidence of a trade-off between workers' compensation and nonwage benefits. The *WCRATE* coefficient is negative, but is not estimated very precisely. It is somewhat surprising that a significant relationship did not surface because such fringe benefits as liberal sick leave, pensions with disability provisions, and various types of insurance would appear to be close substitutes for workers' compensation benefits. But the *WCRATE* coefficient in the full wage equation increases only slightly. Note that this estimate is larger than is implied by theory. My expectation is that it be less than −0.01. This is because the rates apply only to the wage payroll, not to total costs of employee compensation. Consider a worker with the mean hourly wage of $5.81 and hourly nonwage benefits of $0.92. An increase in the workers' compensation rate of $1.00 increases costs by 1 percent of the wage, or $0.058 per hour. For a complete trade-off, total compensation per hour should decline $0.058, or 0.86 percent for the mean worker ($0.058/$6.73).

A possible explanation for the strong result is that variations in workers' compensation costs may be correlated with other mandatory benefits, such as unemployment insurance. Perhaps state legislatures that provide liberal benefits of one type are likely to do so for the other. If these also are negatively related to compensation, the *WCRATE* coefficient would pick up this trade-off as well. The data provide some evidence that this is the case. When *UI,* the firms' unemployment insurance costs per $100 of payroll, was added to the full wage equation, the workers' compensation estimate declined to −0.0098, which is close to the theoretical prediction. The estimate of *UI* of −0.0947 is, however, implausibly high.

One final point concerns the effect of inclusion of the workers' compensation variable in the estimates of compensating differentials for employment hazards. Dorsey and Walzer estimated that omission of the liability variable caused a slight downward bias in estimates of wage premiums. When *WCRATE* was added in this sample, however, the estimates were essentially unchanged. The coefficients on frequency of injury rise, but very modestly.

Conclusions

Empirical evidence suggesting that workers who accept more risky employment receive a compensating differential is both broadened and deepened by these results. Premiums related to increased risk are found in wage and nonwage compensation. Indeed, the data suggest that nonwage compensation is more responsive to variations in employment hazards. The *percentage* compensating wage premium, however, only slightly understates the full differential. Nevertheless, the absolute premium would be understated by focusing solely on wages, and this would be important for estimates of the value of reduced risk. For example, the yearly wage premium for the typical worker with the mean wage would be $395 (0.034 · $5.81 · 2,000 hours) in compensation for a 1 percent increase in the probability of injury. This understates by a substantial margin the value to the firm, in terms of lower total compensation payments, of reducing this risk by one unit: $498.02 = .037 · $6.73 · 2,000.

The evidence is deepened by these results from a different data set and perspective. Because confidence in empirical findings is increased by testing theories in as many different ways as possible, it is significant that these coefficients, estimated from data on firm expenditures, are reasonably close to those obtained from comparisons of individual earnings.

There is now a substantial body of empirical evidence that is consistent with the competitive hypothesis of compensating differentials. This suggests that public policy, which is based upon the assumption that workers are uninformed about work hazards, may have undesirable consequences. For example, regulating safer work environments may make workers who prefer higher risk and higher wage combinations worse off. These findings also imply that firms do have an economic incentive to reduce physical risk.

The issue of compensating differentials is crucial for the analysis of workers' compensation. If risk premiums do exist, increasing workers' compensation benefits will lower wages or other benefits. Estimates from employee compensation data reveal such a trade-off, confirming my earlier finding using employee earnings data. The existence of a complete trade-off does not, of course, imply that workers' compensation is irrelevant. It means that workers are forced to trade the certainty of a small earnings premium for the possibility of a large payment in the event of a workplace injury. In effect, workers' compensation forces workers to buy insurance. Whether this is good or bad policy depends upon one's viewpoint. It can be argued that workers who are less averse to risk would prefer higher risk and compensation. Yet others might maintain that, owing to the small probability of injury, workers undervalue the utility gain from insurance. Other points supporting a mandatory program are economies of administrative costs and preventing free riders who would exploit the income support and medical care available to low-income workers.

These, however, are more modest claims than could be made for workers' compensation if markets failed to compensate workers for risk. The case for this program is weaker if there is evidence that well-informed workers voluntarily choose jobs for which they receive extra compensation. Therefore, a well-documented trade-off may dampen enthusiasm for raising workers' compensation benefits or, at the least, more narrowly define the issues.

The trade-off between earnings and workers' compensation also suggests that recent concern in some states over possible employment losses due to increased workers' compensation costs may be unfounded. If firms can offset these costs, there will be no employment disincentives.

5 · COMPENSATING STRESS-INDUCED DISABILITY: INCENTIVE PROBLEMS

Michael Staten and John Umbeck

The rising costs of administering workers' compensation programs have been a chief concern of insurers in recent years. Expansion of benefit coverage to encompass occupational diseases is perhaps the most significant contributing cause. Occupational diseases have been implicitly covered for many years, but advances in medical technology are revealing with increasing frequency new links between the work environment and medical disorders. For insurers, this poses acute difficulties in computing premiums and equitably distributing burgeoning costs across the system.

One recent trend has been approval of more stress-induced disability claims as occupational diseases. Work stress occurs in a variety of forms and spawns a variety of physical symptoms. Regardless of the manifestation of stress, the problem of distinguishing its origin

---

For financial support, the authors are indebted to the Carthage Foundation, the Krannert Graduate School at Purdue University, and the University of Delaware. We also thank John Barron, James Chelius, Robert Clower, Fred Siskind, and an anonymous referee for helpful comments on earlier drafts. Portions of the following discussion and tables previously appeared in Staten and Umbeck, "Information Costs and Incentives to Shirk: Disability Compensation of Air Traffic Controllers," copyrighted (1982) by the American Economic Association and reproduced with permission of *The American Economic Review* — M.S. and J.U.

— work-related versus non-work-related — is formidable. For certain types of disorders, most notably psychological and emotional reactions, it is difficult to accurately document and measure the *existence* of the problem, a determination more fundamental than causality. These disorders are not always accompanied by physical symptoms. In assessing this type of claim, a physician, to some extent, must rely on the employee's own statement that he or she has a problem. This study addresses the difficulties of determining the existence and cause for claims of mental stress.

Mental Disability and Compensation Law

Although the various state compensation acts uniformly cover the majority of physical injuries that arise on the job, coverage of psychological disabilities is diverse. The infinite variety of stimuli that can induce emotional reactions probably contributes to the lack of standardized treatment. A three-part categorization of these cases has been suggested to clarify some of the adjudication problems (Larson 1970). *Mental-physical* cases are those in which some nonphysical stimulus induces a physical response. Examples of workers witnessing frightening events and suffering heart failure shortly thereafter abound. A case of prolonged anxiety or stress (even for periods of weeks or months) that causes, say, an ulcer, heart failure, or stroke also falls in this category. Whether the stressful incident is sudden or prolonged, the distinguishing factor is the resulting physical ailment that can be readily documented. *Physical-mental* cases fall in the second category. These are claims of neurotic disorders that follow some physical injury. Continued paralysis of a limb, psychologically induced, after a standard healing period is an example.

The final category of cases is *mental-mental* claims, in which some nonphysical event, sudden or gradual, triggers a psychological reaction that disables the worker. Perhaps a worker witnesses an explosion at a plant and, although physically uninjured, is unable to return to work without some neurotic symptoms. The worker is the victim of no physical injury but is, nevertheless, incapacitated.

Historically, state jurisdictions have readily awarded compensation in cases in which some physical injury is easily documented.

Reluctance has come where readily detectable injury was absent. Thus, the first two categories of cases are generally compensable, with the states divided in their treatment of cases in the third category. Larson (1970) has suggested that legal attitudes have been analogous to the sentiment that certain types of physical discomfort are "all in one's head."

> "How could it be real when it was purely mental?" This poignant judicial cry out of the past which I occasionally quote to put down psychiatrist friends, contains the clue to almost all of the trouble that has attended the development of workmen's compensation law related to mental and nervous injuries. This equation of "mental" with "unreal" or imaginary or phoney is so ingrained that it has achieved a firm place in our idiomatic language (1970, p. 1243).

Obviously, a phony disability deserves no compensation. The question that has plagued compensation appeals boards is whether mental-mental cases can ever be real disabilities.

Legal precedent in some states suggests that the law does not recognize the existence of a compensable disability in certain classes of mental-mental cases. Two recent discussions in legal journals focus on this issue (Gross 1977; Filteau 1980).[1] The Massachusetts Supreme Judicial Court has raised a distinction between mental-mental injuries induced by sudden stress and those created by gradual stress. The rationale seems to lie in the court's attempt to set guidelines for meeting the work-related test required by law. Medical science can rarely trace a relation between a mental disorder and a unique cause. Generally, medical testimony will list several probable causes, leaving the claimant's occupation as a possible but not certain source. Thus, the work-related test for compensation is nearly impossible to satisfy with certainty. To be sure, Massachusetts case law makes it clear that a compensable mental-mental injury can occur, but the burden of proof that falls on the claimant is compounded by the possibility of nonwork sources of stress. Although it helps to show evidence of good work experience and sound mental health before the disability, it remains difficult to firmly establish cause and effect.

---

1. The following discussion borrows heavily from their summaries of Massachusetts and New York case law.

To facilitate the forging of the causal link, the courts have employed a practical if not entirely equitable guideline. Current case law governing disposition of mental-mental cases requires an examination of the nature of the stressful event. Massachusetts courts have resorted to using the *duration* of the stimulus precipitating the disability as a proxy for whether the disability was work-related. A well-documented traumatic experience *at work* (such as an explosion or the death of a coworker) before the onset of symptoms strongly suggests cause and effect. Compensation may well follow. Conversely, if the claimant succumbs gradually to continued stress at work, the longer duration of the stimulus suggests a much weaker link between work and the disability; the prolonged development of symptoms allows more opportunity for nonwork stress to cause the problem. Compensation is unlikely, with some version of a normal-wear-and-tear argument cited as grounds for denial.[2] Thus, even if the mental injury is truly connected to the workplace, the decision turns on the duration of the stimulus causing it.

Massachusetts is not unique in the practical implications of its compensation law. New York draws a similar distinction between gradually induced disorders and injuries that result suddenly from an event. Simply put, New York law requires that an accidental injury be either the result of a sudden, unexpected event or be manifested in the sudden onset of a disability (such as heart failure). A sudden injury occurs at a given time in a specific place. Incorporating this feature into a working definition of accidental injury lends itself to procedural convenience by answering a variety of questions necessary for determining compensability (e.g., work relation, timeliness of filing). Of course, only accidental injuries are compensable under New York law. So, practical application of the law by the courts has incorporated time and place requirements into the test for compensability. Both mental-physical and physical-mental cases satisfy the requirement that they have a sudden cause or sudden, observable effect. Lacking

---

2. Earlier this century, Massachusetts courts occasionally relied upon a wear-and-tear doctrine to justify denials of claims for gradually developed physical disorders. The idea was that years of prolonged activity in occupations requiring physical labor simply caused the body to wear out. Thus, years of living, rather than the job, per se, caused the disability. The principle was usually applied to cases in which no single disabling event occurred.

either, mental-mental cases of prolonged stress and gradual psychological breakdown presumably would fail a test as accidental injuries.

Assuming that prolonged stress can, in fact, induce psychological disability, these disorders presumably fall under the scope of the workers' compensation statutes. Both Massachusetts and New York (to cite but two examples) have apparently adopted rule-of-thumb proxies for sorting mental-mental cases. The reliance on duration of stress as a proxy for work relation discriminates against genuine but gradually induced work injuries. But perfect information on causality rarely exists. Its absence forces some administrative approximations. These practices carry the implicit suggestion that the social loss from erroneously compensating nonwork injuries is deemed greater than the loss from denying benefits to eligible victims of gradual stress.

States have expressed equal or greater concern over compensation fraud in mental cases (Lasky 1980). Establishing the existence of a genuine mental disability presents a formidable task. In gradually induced mental-mental cases the physician must, to some extent, rely on the claimant's own statement that he or she has a problem.

From a contractual standpoint, the information problem creates incentives for abuse. Workers' compensation is a contingent claims contract, as are all insurance agreements. Payment of benefits is contingent upon the occurrence of some prespecified event, in this case a work-related injury or disease. If the existence of the event upon which payment is contingent is difficult to determine, incentives to exaggerate or falsify claims are created. This results from the possibly significant probability of avoiding detection and capturing benefits.

Distinctions between sudden versus gradual onset in stress cases reduce the threat of fraudulent filing. The requirements for specific time and place provide a reference point even if the disorder itself lacks visible symptoms. Use of the time and place distinction suggests a judicial belief that mental disability may legitimately follow some traumatic shock but is imagined or faked in its absence. However, this practice may compound the incentive problem it is designed to combat. A worker who can be compensated only in the case of a traumatic event has an incentive to provide one; insistence on time and place documentation may well encourage more accidents. Nevertheless, the requirements are very useful tools for processing notoriously difficult cases.

The problems posed by mental-mental cases are somewhat alleviated by the small number of cases presented in most states. But for one group of employees covered by the Federal Employees' Compensation Act (FECA) the incidence of psychological disability is high and the claim volume significant. Air traffic controllers are the largest single group of employees routinely subjected to mental stress and specifically covered for mental disorder by a compensation system. As such, they offer a unique opportunity to examine problems and incentives created by compensating stress-induced disability.

Disability Programs for Air Traffic Controllers

With an average work force of about twenty thousand employees during the last decade, air traffic controllers are the largest single occupational group within both the Federal Aviation Administration (FAA) and the larger U.S. Department of Transportation. In return for providing a crucial service to the flying public, they are among the highest paid federal employees. They have also experienced medical disqualification and subsequent retirement at rates higher than other federal occupations. The notion that controlling aircraft is physically and mentally taxing has been generally accepted over the last decade, although debate over claims of excessive stress lingers. Medical records reveal the most common controller complaints are stress-related: psychological trauma, heart disease, and ulcers.

This presents two problems for FAA officials. On the one hand, maintaining proficiency of the controller work force is of paramount importance. Controllers medically unable to perform and those likely to develop disabilities under continuing stress must be eased from their sensitive positions. But if the FAA gains a reputation for dismissing controllers at the first sign of a problem, qualified controllers could become difficult to recruit. Controllers make heavy investments in specialized training (two to five years) before they can assume full duty. There are few opportunities outside the FAA for employing these skills, particularly at the salary of an experienced controller.

This dilemma is in part why controllers were covered by two types of disability and retraining programs from 1972 to 1978, coverage unique to their occupation.

Regardless of the state in which they reside, all civilian federal employees have disability insurance under the Federal Employees' Compensation Act (FECA). Originally passed in 1917, FECA has been administered by various bureaus within the U.S. Department of Labor. A major reorganization in 1974 produced the Office of Workers' Compensation Programs, OWCP, which currently handles the task. Under the provisions of the act, compensation takes the form of either a fixed award or a percentage of salary paid for the duration of the injury. The nature of the injury determines the type of payment. Unlike state-administered programs in which benefits are defined in terms of statewide average wages, FECA payments are tied directly to the claimant's salary. Employees without dependents who qualify for salary percentages receive tax-free 66.66 percent of the base rate salary at their GS level. Employees who have dependents receive tax-free 75 percent of their base salary. Compensation duration is a function of the employee's recovery rate and ability to resume his or her original job. Payments cease when OWCP determines that an employee has fully recovered. A worker only partially disabled is not always entitled to the full percentage payment of salary; payments may be reduced after some initial recovery period, depending on OWCP's assessment of the employee's "wage earning capacity." This provision was designed to encourage disabled workers to return to limited duty when possible. The program provides up to forty-eight months of rehabilitation, retraining for another occupation, or both if return to the original job is impossible. Finally, OWCP pays all medical bills related to the injury, as well as travel expenses connected with treatments. Of course, to qualify for compensation all injuries must be verified as work-related. FECA, however, differs significantly from most state programs in its nonadversarial nature. An employee can appeal an adverse decision made by the OWCP examiner. The employing agency cannot.

In May 1972, Congress authorized funding for the ATC Second Career Training Program. It was designed to provide training for a second career to controllers removed from active duty because the employee was medically disqualified for duties as a controller or such removal was necessary for the preservation of the physical or mental health of the employee. The program did not require that the cause of the problem be job-related. If such a controller had been employed at least five years, the FAA agreed to pay for an approved training program of the controller's choice. Counseling and vocational testing services were provided. The FAA required the employee to complete the program within three calendar years from entrance and paid the base rate for up to two years from the date of disqualification.

The lack of a work-related test for eligibility was the key to the program's solution to the safety-recruitment dilemma. FECA benefits were, in principle, always available to the victims of work-induced stress. The FAA perceived that more controllers were developing disorders than were filing notice of disability. One reason controllers were not filing was fear that FECA benefits would be denied due to lack of concrete evidence of either the disability itself or the work connections. A controller who risked an adverse decision on compensation by filing an injury notice also disqualified himself or herself from duty. The requirements of the retraining program allowed the controller to reveal a problem without risking an immediate loss of income. Similarly, a person could undertake a career as a controller knowing that, should medical reasons prevent him or her from remaining in the position, there were viable alternatives to prevent a substantial reduction in income.

The retraining program admitted entrants from May 1972 until October 1978. The FAA's experience of completion rates was termed "disappointing" in a 1978 General Accounting Office study that recommended the program be discontinued. The program's structure, the nature of the disabilities covered, and the ability of controllers to use both retraining and FECA jointly contribute to an interesting pattern of injury reporting through the 1970s.

Incentives to Report Injuries

All types of controller disabilities, psychological or otherwise, are covered under FECA and the retraining program. Controllers suffer a high incidence of mental-physical and mental-mental disorders, while physical-mental injuries are relatively rare, probably because of the sedentary nature of the occupation. We focus on mental stress injuries because the inherent problems in identifying them contribute to the filing incentives discussed below.

It may be useful to illustrate the categories of mental claims in terms of controller experience. Suppose a controller witnesses a near mid-air collision, or worse still, a fatal crash. The resulting nausea, gastritis, hypertension, or heart attack would fall into the mental-physical category. Similarly, constant work pressures, without the traumatic incident, that create the same maladies also create mental-physical cases. If the gradual work pressure or the fatal crash induces an emotional or psychological reaction (such as insomnia, paranoia, or schizophrenia) the disability is mental-mental. Examples of each are readily documented in OWCP files. Determining causality presents the chief obstacle to adjudicating cases in the first category since job stress may have been only a secondary cause. Of course, cases of heart failure resulting from gradual stress are trickier to prove than heart failure after a crash. But for mental-mental cases both causality and existence are difficult to assess, although a frightening incident seems to reduce initial doubt.

Truth and Consequences: Reporting Genuine Disabilities. A recurrent theme in debates over compensability of mental disorders is the concern that they are too easily faked. But the sword of uncertainty cuts both ways. If it is difficult to detect when an employee fakes mental disability, it may also be hard to tell when he or she is disguising a genuine problem.

Consider now a controller with a genuine medical problem, stress-induced but not necessarily mental, before the 1972 enactment of the second career program. FAA regulations mandate that every controller must submit to an annual physical examination. If symptoms were found of a potentially work-impairing disorder, the controller could be disqualified from the job. If the costs of enduring the ailment were lower than the costs of taking other employment (probably at a lower salary because a controller's specialized skills aren't easily transferable), the controller would have some incentive to conceal the disability to avoid being dismissed. It seems reasonable to suppose that the most difficult problems to measure are the easiest to mask. Ulcers can be detected with X rays. Hypertension and irregular heartbeat are also readily documented. But, emotional and psychological problems may escape detection. Therefore, this argument suggests these disorders are successfully hidden by controllers more often than any other class of disorders.[3]

The second career program substantially lowered the cost of revealing a disorder and taking other employment. We would predict an increase in controller dismissals for a variety of medical reasons after May 1972, but a more than proportional increase in emotional and psychological disqualifications. Of course, a controller with a genuine problem was entitled to FECA benefits if he or she could show it was caused by the work. Once the controller revealed the problem, the additional cost to him or her of filing for compensation was relatively low. Hence, disability claims should also have increased after creation of the program.

Compensation Fraud: "Punching Out." Potential fraud always exists when insurance coverage is offered in a world of costly information. The problem is not unique to workers' compensation. However, the generosity of FECA benefits for higher paid federal employees such as air traffic controllers creates an incentive for manufacturing a claim.

Assume that an insurer must incur costs in order to verify the existence of an alleged disability. Maximizing insurers will gather information as long as the gains exceed the costs at the margin. Assuming diminishing marginal value of information and increasing marginal costs, the implied equilibrium will leave some information at large. In other words, some uncertainty will remain concerning the true nature of the loss claimed by the worker. This raises the possibility that benefits might be paid without a loss actually occurring or that benefits paid will exceed the value of the loss. To the extent that this is promoted by efforts of the worker, attempted fraud exists.[4]

Clearly, the equitable operation of the insurance contract's payoff mechanism depends upon the effectiveness of the monitoring undertaken, a function of the cost of gathering relevant information. For this reason, the insurer's cost of monitoring plays a crucial role in an individual's decision to file for benefits. Consider the decision process of a utility maximizing worker filing a fraudulent claim. There is

3. This argument implicitly assumes that a successful OWCP claim upon disqualification is no certainty.

4. Of course, there is also the possibility that the insurer may erroneously deny compensation when it is deserved.

probably a variety of injuries or disabilities the worker can allege in order to receive a given benefit package. The worker must decide which alternative will yield the highest expected utility, assuming only one may be alleged. Next, he or she must determine how much to spend on faking a loss, so as to maximize the utility within the chosen alternative.

A model of this decision process will only be summarized here.[5] First, it is important to note that the insurer's monitoring costs and the worker's efforts to obtain a given level of benefits are interrelated in the sense that one's behavior influences the cost of the other's effort. For example, suppose the insurer hires additional claim examiners, so that each examiner spends more time investigating each case. This additional monitoring requires the worker to increase his or her own efforts at manufacturing a claim to counteract the now greater probability of detection. Conversely, as monitoring declines, workers find the process of faking a claim with any given probability of success becomes cheaper. A model can be used to show that two separate events, an increase in compensation benefits and an exogenous increase in the insurer's cost of monitoring claims, will have similar effects. They each result in a rise in the *probability* of successful fraud as well as an increase in the *filing* of fraudulent claims.

Now consider FECA and controller incentives for fraud. On average, a disabled controller with at least one dependent qualifies for tax-free disability compensation that actually exceeds normal take-home pay. Since a controller receives compensation for the duration of the disability, upon presenting a convincing case he or she could make more by staying on the compensation rolls than by staying on the job.

Several factors, including the FECA administrative changes in 1974, contributed to put the controllers in a unique position to take advantage of the program. The types of disabilities controllers have been acknowledged to suffer are the very ones posing the greatest processing difficulties for the compensation system. Throughout most of the past decade, an OWCP claim examiner had fewer *a priori* doubts about a controller claim of psychological stress than about a similar claim from, say, a federal office worker. Still, such a claim is

5. Details are available from the authors upon request.

difficult to assess. In particular, the degree and duration of the disorder are difficult to project, since behavioral symptoms may be recurring and such disorders are associated with no standard healing time.

After the FECA amendments of September 1974, the testimony of clinical psychologists was permitted for the first time as primary medical evidence of a disability. More important, employees were allowed to select the private physician of their own choice to supply medical testimony for their claims (5 U.S.C. § 8101 (1976)). Before 1974, verification of a controller's disability was usually made by an FAA flight surgeon or, on occasion, by some designated private physician.[6] We do not want to suggest that FAA physicians were never sympathetic to controller disabilities. But because the doctor was selected by the government, an employee's ability to shop around for a sympathetic physician was limited. The change allowed greater freedom to find a doctor sympathetic to a claim or one who might cooperate for a price. In theory, by reducing the effectiveness of each monitoring dollar, these changes would have raised, to some degree, OWCP's cost of monitoring the veracity of *all* employee claims. But the effectiveness of monitoring dollars would have been most reduced for claims of stress-induced psychological and emotional problems, which were already medically difficult to assess.

It is significant that these program changes came during a period in which OWCP abolished its investigative staff. The task of conducting follow-up investigations for long-term recipients was assigned to claim examiners already experiencing increased caseloads. In addition, OWCP failed to increase the size of its staff of examiners proportionally with claims filed and did not automate or otherwise make technological improvements to augment processing productivity (U.S., Congress, House 1976).[7] Apparently, the government

6. An exception to this statement should be noted. Employees had formerly been required to use federal physicians *when available*. If federal medical facilities were not available, employees had the right to consult physicians designated and approved by the U.S. Department of Labor. In 1972 the department effectively approved all private, licensed physicians. Few controllers, however, would have been able to use nonfederal physicians since most FAA facilities are in populated areas or have flight surgeons on the staff.

7. From 1970 to 1976, while the OWCP staff increased 36.6 percent, injury reports, new compensation claims, and continuing disability claims increased 58.5 percent, 126.6 percent, and 80.7 percent, respectively.

chose not to increase monitoring at a time when existing efforts were losing effectiveness.

"Punching out" is industry jargon for the process of alleging a medical disorder, being disqualified from duty, and successfully landing a spot on OWCP's periodic roll of long-term compensation recipients. Psychological disorders emerge as likely candidates for fraudulent claims because of their stress-related nature, the government's difficulty in verifying their existence, and the fact that they offer the greatest opportunity for long-term compensation since they are characterized by no standard recovery period. Given the prevailing attitude toward stress as an occupational hazard and the controller's relatively easier task of obtaining medical support for a psychological claim after the amendments, we predict a more than proportional increase in the incidence of successful fraud alleging psychological disability.[8]

This is not to suggest that there were never fraudulent claims of other disorders. But the amendments not only increased the government's monitoring costs, but also reduced the relative cost of claiming psychiatric disorders compared with other diseases. In other words, for those controllers who may previously have had a comparative advantage in faking some other disease, the shift in relative costs would cause some of them to switch to psychological disability claims. Thus, the greatest change caused by the amended filing requirements should have appeared in the psychological category of claims.

Mental-Mental Claims and Job Performance. Like several states that have required employees seeking compensation for mental-mental claims to specify a time and place of injury, OWCP also adopted a similar rule for adjudicating claims of emotional stress. In an effort to formalize the criteria for approving stress-induced psychological claims, OWCP examiners were instructed to look for specific stressful incidents on the job that were symptoms of or contributed to a disability. Controllers were asked to provide written statements of the events

8. The existence of the second career program compounded the incentives to fake a disability. In one sense, the program could be viewed as a backup in the event a claim was denied. Perhaps more important, the program provided a continuation of income to a controller between the date of dismissal and the date the compensation claim was approved, a lag that often exceeded six months. The retraining program did not prohibit controllers from quitting upon approval of a disability claim.

TABLE 5.1
Changes in Disease Incidence among Controllers

Disorder	Incidence Rates		Percentage Change
	Pre-Second Career	Post-Second Career	
Eye	5.6	7.4	+32.0%
Ear, nose, throat	6.7	8.0	+19.4
Respiratory	1.9	1.5	−21.0
Cardiovascular	22.1	32.5	+47.0
Abdominal	16.7	20.4	+22.0
Neuropsychiatric	10.9	27.2	+150.0
Bones and joints	2.3	4.6	+39.0
Muscles	0.5	0.4	−20.0

Source: These figures are based on data reported in *FAA-Aviation Medicine Report,* No. 78-21.

Note: Incidence rates were computed per thousand person-years of service to adjust for varied entry and exit dates of the work force. The pre- and post-second career periods are May 1967 to May 1972 and May 1972 to May 1977, respectively.

they believed contributed to the problem. This testimony was corroborated by the official statement of a controller's supervisor, required to accompany a claim.

This point deserves emphasis since it raises the remarkable possibility of a specific link between the compensation program and job performance. To obtain approval for a claim of emotional or psychological stress controllers needed evidence of noticeable deterioration in job performance over a period of time. This suggests an incentive for a controller filing a claim to either allow his performance to noticeably deteriorate or to cite evidence of deterioration if it already exists. The requirement seemed designed more to weed out exaggerated or fabricated claims than to discriminate against victims of gradual stress. OWCP records show that compensation was granted to victims of "sudden stress" (e.g., a controller who witnessed a crash) and to those developing symptoms gradually. The incidents cited by the controllers apparently lent authenticity to both types of claims.

A unique aspect of an air traffic controller's occupation is that the job is essentially one-dimensional — the task is to maintain the separation between aircraft in the assigned airspace. A controller's performance of this task is more closely monitored than the work of perhaps any other federal employee. Separation violations, when detected, are carefully documented and investigated. This suggests that any controller filing a stress-related psychological claim, either real or fabricated, had an incentive to supply a separation violation as evidence. He or she could claim either that the event itself was so traumatic that it caused mental stress or that the job caused the stress that impaired his or her ability to function, as evidenced by the incident. Either way, the incident would serve to establish the claim as genuine. Therefore, we would expect reports of such violations to increase with every increase in the incentives for filing such claims.

Filing Experience

By lowering the expected costs of claiming a disability, we would predict that the second career program and the 1974 FECA amendments would each lead to an increase in disability claims filed, some genuine and some not, and to a more than proportional increase in psychological and emotional claims, some genuine and some not.

The most consistent measure of controller health available is the summary data from the annual controller physical examinations required by law. The FAA's Office of Aviation Medicine has conducted a study of 25,517 controllers who received one or more medical examinations between 1967 and 1977. Table 5.1 contains disease incidence rates from the study. These rates were computed (per thousand person-years) for five-year periods before and after the advent of the second career program in 1972. Incidence rates for most classes of disorders rose, but the most striking increase was for psychological disease, as predicted. Unfortunately, the FAA data did not allow subdivision of exams in the post–second career period to isolate the separate effect of the FECA amendments.

Higher disease incidence rates after the passage of the retraining program should have spawned more disqualifications. Tables 5.2 to 5.4 summarize controller disqualifications from fiscal year 1973 through 1978. Fiscal year 1973 began on July 1, 1972, coinciding with the start of the retraining program. The statistics in table 5.2 are divided into two categories based on eligibility for the FAA's early retirement program (age or length of service). The volume of disqualifications rose steadily through the period before peaking be-

TABLE 5.2
Medical Disqualifications among Controllers

	Number not Otherwise Eligible for Early Retirement	Also Eligible for Early Retirement	Total
1973	143	56	199
1974	259	95	354
1975	354	115	469
1976	450	96	546
Transition quarter	179	27	206
1977	364	68	432
1978	263	61	324
10/1/78–12/31/78	36	6	42
Total	2,048	526	2,574

Source: Summary statistics were provided by the FAA.

Notes: All years cited are fiscal years. For the period 1972–76, a fiscal year ran from July 1 to June 30. In 1976 the government underwent a transition quarter during the switch to an October 1 to September 30 fiscal year, which began with fiscal year 1977.

*The second career training program ceased accepting new entrants as of October 1, 1978. This column reflects disqualifications in the following quarter.

TABLE 5.3
Reasons for Controller Removal

Medical Disorder	Source of Original Examination				
	Private Physician	FAA Regional Flight Surgeon	Aviation Medical Examiner*	Total	As % of Total
Psychological	586	429	48	1,063	52.0%
Cardiovascular	192	143	21	356	17.4
Gastrointestinal	111	34	5	150	7.3
Defective hearing	43	85	23	151	7.3
Arthritis-skeletal	23	17	1	41	2.0
Defective vision	8	18	2	28	1.4
Other	149	103	6	258	12.6
Totals	1,112	829	106	2,047	

Source: Summary statistics derived from an FAA survey.

Notes: These statistics pertain to controllers disqualified from duty from July 1, 1972, through March 31, 1977.

*Annual controller medical exams are occasionally contracted to FAA-designated aviation medical examiners not subject to controller choice.

TABLE 5.4
Age and Length of Service of Disqualified Controllers

	Number Removed	Percent of Total
Age		
25–29	18	0.9%
30–34	243	11.9
35–39	267	13.0
40–44	398	19.5
45–49	332	16.2
50–54	344	16.8
55 and over	445	21.7
Total	2,047	
Length of Service		
5 years	63	3.1%
6 years	195	9.5
7 years	133	6.5
8 years	87	4.2
9–10 years	92	4.5
11–14 years	236	11.5
15–19 years	744	36.4
20–24 years	280	13.7
25–29 years	176	8.6
30 or more years	41	2.0
Total	2,047	

Source: Statistics derived from an FAA survey.

Notes: These statistics pertain to controllers disqualified from July 1, 1972, through March 31, 1977.
Years of service include initial training. Five years of service were required for second career retraining eligibility.

tween mid-1975 and mid-1976. Volume declined thereafter, with a sharp decline in the number disqualified during the fourth quarter of 1978. The retraining program froze admissions as of October 1, 1978, eliminating paid retraining as an alternative.

Table 5.3 cites the medical reasons for disqualification for 2,047 controllers removed from duty from July 1, 1972, through March 31, 1977, under the auspices of the retraining program. Fifty-two percent of all disorders precipitating removal were originally diagnosed by private physicians chosen by the controllers. A few received examinations from private physicians designated as aviation medical examiners by the FAA. The FAA contracts with the latter group to do annual physical examinations. They are not chosen by controllers.

Age and length of service are summarized in table 5.4. Again, five years of experience, including training, were a prerequisite for retraining. Most disqualified controllers (nearly 74 percent) were over forty years of age, and over half were veterans of at least fifteen years of service. At first glance, these figures do not seem inconsistent with the program's intent to maintain a young, proficient work force. An additional 28 percent of disqualified controllers had between five and ten years experience. FAA personnel statistics reveal that the proportion of the total work force with five to ten years experience averaged between 25 percent and 29 percent for roughly the same period. But the fundamental premise of the retraining program was that the risk of burnout and medical disqualification *increased* with years of service. If the premise is correct, then young controllers should represent a smaller proportion of the medically disqualified group than of the total work force. The observed proportion of five- to ten-year veterans is not consistent with the unmasking of genuine disorders argument as the sole explanation for the increase in disqualifications. It is consistent with the punching out argument.

A controller who is considering punching out of the system must choose the point in his or her career to file a claim. Of course, by doing so the controller sacrifices any future salary increases. We would expect a maximizing controller to punch out at that time when the expected gains from waiting another period have fallen to a level that just equals the value of the leisure relinquished by working one more period. The gains can be roughly calculated using the present value of the benefit stream a controller will receive over the life of the disability, a stream that is tied to salary. We will assume the controller's leisure value is constant.

Disability benefits are a tax-free percentage of the controller's highest salary. Controllers begin employment (as trainees) at a level of GS-7. Normal training progress allows them to advance to a GS-9 after one year and to a GS-10 or GS-11 after two. After these two increases they can move up a maximum of one level a year until, after their fifth year, those at larger facilities attain GS-15 status, when advancement ceases. At 1982 wages, controllers can move from about $15,000 to over $45,000 in five years. Subsequently, their incomes rise relatively slowly.

A controller who punches out during the first five years incurs a substantial opportunity cost in terms of the present value of sacrificed salary increases. This opportunity cost is much smaller if the controller remains employed through the first five years of rapid advancement. Five years' service are also required for eligibility in the second career program, which could be used as a backup and income supplement between the dates of disqualification from duty and claim approval. After the fifth year of employment the gains associated with waiting an additional year fall such that we would predict some punching out would begin. The observed proportion of disqualified five- to ten-year veterans is by no means conclusive evidence, but it is certainly consistent with a pattern suggested by a theory of compensation fraud.

Next, consider data from OWCP on injury reports submitted by controllers (table 5.5). An OWCP examiner creates a case file for each controller injury report submitted. The examiner gathers medical testimony and statements from the injured controller, supervisors, and other persons witnessing the accident. A decision on eligibility is reached, and the case is classified according to the type of benefits authorized. Some injuries with relatively short recovery periods are authorized for medical care payments only. These controllers may elect to take accrued vacation leave to recover rather than filing for compensation. Claimants with more serious injuries and longer recovery periods are placed on the periodic roll, the list of recipients of regular compensation payments. The 1974 FECA amendments introduced still another category of claims. Continuation of Pay (COP) benefits apply to injuries with recovery periods of forty-five days or less. Injured employees who do not take sick leave continue to draw regular pay from their employing agency. The provision was designed to protect the worker from loss of income between the date of injury and OWCP's final decision on compensation, a decision that occasionally took several months.

Although OWCP records are useful for determining the volume of injury reports filed, they have a serious drawback in the large proportion classified as closed, retired, or destroyed. When OWCP

TABLE 5.5
Status of
Controller Injury Reports,
1969–79

Case Status	As Percentage of Total
Developmental stage	2.4%
Death roll	0.8
Medical care only	1.6
Periodic roll	8.0
Leave taken	2.0
Continuation of pay*	3.3
No lost time	17.3
Compensation denied	3.0
Closed, decision unknown	14.0
Retired and destroyed	45.0
Other	2.6
Total	100.0

Source: Compiled from OWCP data.

Notes: This table summarizes case status of April 1, 1981, for more than nine thousand cases created from injury reports filed during the period 1969–79.

*Under the 1974 FECA amendments, controllers with short-term recovery periods (45 days or less) received a continuation of their pay directly from the FAA.

files were automated, much of the specific information on older cases was omitted from the coded records. Without inspecting each of the individual case files, there is no way of knowing the final disposition of nearly 60 percent of the injuries reported. Thus, the table understates the proportion of injuries falling in each of the payment categories, including injuries resulting in no lost time. The percentages presented in table 5.5 reflect case status as of April 1, 1981. Eight percent of the injuries were serious enough to still be compensated in 1981. Of course, many more controllers may have been placed on the periodic roll and later removed upon recovery. Data on these kinds of claims would be embedded in the closed or retired category. The 767 controllers remaining on the active rolls are summarized in table 5.6.

TABLE 5.6
Periodic Roll Recipients
by Type of Disability

	Occupational Disease						
	Cardio-vascular	Psycho-logical	Gastro-intestinal	Other Vascular	Traumatic* Injury	Other and Unknown	Total
1969	3	7	3	1	6	2	22
1970	6	34	8	9	7	9	73
1971	14	32	6	7	16	6	81
1972	13	54	7	9	23	12	118
1973	18	68	14	16	30	20	166
1974	14	53	7	7	18	19	118
1975	13	18	4	3	18	4	60
1976	7	31	3	5	17	1	64
1977	5	9	0	2	16	0	32
1978	2	4	0	1	10	2	19
1979	2	1	1	0	10	0	14
Total	97	311	53	60	171	75	767
Total as percentage of all cases	12.6%	40.5%	6.9%	7.9%	22.3%	9.8%	

Source: Compiled from OWCP data.

Notes: Case types are for all recipients of monthly compensation payments as of April 1, 1981.

*Includes all types of lacerations, contusions, fractures, sprains, hearing loss, etc.

Disorders that OWCP classifies as occupational diseases comprise about 65 percent of the claims.[9] In particular, 311 controllers (40.5 percent of the total) were receiving long-term benefits for psychological disorders. The percentages are generally consistent with the breakdown of disqualified controllers in table 5.3.

The most striking aspect of the statistics is that more of these *active* cases were created in 1973 than in any other year, before or since. Again, this does not mean that 1973 produced more periodic roll recipients than any other year. Many long-term cases may have been closed within two or three years after creation. The table applies only to those still receiving benefits. It does, however, suggest that

9. Many controllers suffered several problems at once. For example, a primary psychological disability might also be accompanied by hypertension or an ulcer. Cases in table 5.6 are classified according to the primary medical disorder.

claims filed in 1973 were judged more serious than those in subsequent years. Table 5.2 revealed that 1973 was not the peak year for disqualifications. But, if the controllers removed after 1973 were approved for the periodic roll, on average they did not remain on it as long as their earlier counterparts.

Given the incentives to disguise disorders before 1972, it is reasonable to attribute the increased disease incidence to the unmasking of genuine disorders. This would also explain an increase in medical disqualifications. But the volume of annual disqualifications was still rising three years after implementation of the retraining program. In light of the statistics on length of service, it seems unlikely that the unmasking argument is the sole explanation of the continued increase in dismissals. The evidence also supports the argument that claims continued to rise because of the work disincentives created by making generous compensation payments easier to obtain. The data on OWCP claims filed do not allow much refinement of these conclusions. The lack of detail on how closed claims were adjudicated prevents the tracking of compensation payments over time. We have information only on those controllers still receiving benefits, which certainly understates the number who successfully filed.

To this point, no mention has been made of data on job performance during the period. To what extent did the incentive to provide an incident on the job as evidence of mental stress affect measured performance? A study of performance measures from 1973 to 1976 is detailed in Staten and Umbeck (1982). To summarize, an increase in documented errors on the job appeared simultaneously with the rise in disqualifications and claims. The increase was statistically significant, even after controlling for exogenous factors, such as technology changes and air traffic volume, that might have affected measured performance. Characteristics of both the controllers committing errors and the errors themselves were consistent with predicted trends. One finding is of particular interest. Consistent with the medical disqualification data, controllers experiencing the largest increase in reported errors were five- to ten-year veterans. The incidence of errors in this group is not consistent with the notion that burnout risk (and accompanying performance deterioration) increases with age and years of service. Thus, the unmasking argument does not seem to explain this change in composition of the work force that committed errors.

We are lacking one piece of information crucial for completing a picture of the exact way controller claims of stress were adjudicated. We know that the agency asked controllers to refer to specific incidents, and we also know that more incidents actually occurred. But no data have been available on the percentage of accepted claims for which no incident was cited or the proportion of denied claims that did cite one. In other words, there is no evidence how closely OWCP followed its own guidelines. This information would presumably have become available to controllers as the various FAA installations accumulated filing experience. Over half the controller work force was employed at a relatively small number of facilities (twenty), each employing three hundred to six hundred controllers. Word of mouth and a well-organized union could quickly distribute information about the success rate and characteristics of applicants. This, in turn, could have influenced subsequent claims. A detailed study of the evidence used in controller claims is underway and should provide some insight into these issues.

Conclusions

The FAA has grappled with the incentive problems created by the dual coverage offered in second career retraining and FECA benefits. The agency's solution, imposed in 1978, was to dismantle the training program. Completion rates had been dismal during the first five years of the program, despite an expenditure of $104 million. Of 2,047 controllers eligible for retraining as of March 1977, over 40 percent either declined training or terminated before completion. Only 22 percent had completed their scheduled program as of that date, and a General Accounting Office survey in 1978 estimated that as few as 7 percent of eligible controllers actually pursued the second career for which they trained. The program carried no restrictions prohibiting dropping out upon OWCP claim approval or requiring payment of training costs in the event of dropout. In addition, OWCP also offered rehabilitation and retraining programs under FECA. Of course, access to the FECA retraining required a job-related disorder. In a 1979 report, the FAA rationalized that benefits available to controllers under FECA, Civil Service Disability Retirement (an annuity with no job related injury requirement), and the FAA's own early retirement package were sufficient to eliminate the need for second career training (U.S., FAA 1979).

This still leaves the incentive problems inherent in the FECA payments. Assume that medical technology does not improve to the extent that a psychological disorder becomes as easy to diagnose as a broken leg. OWCP could impose more stringent requirements for evidence necessary for documenting a problem. Claim volume might fall, but experience suggests controller errors on the job would probably rise. Alternatively, suppose that to avoid creating further incentives for performance errors, OWCP were to drop all requirements for specific incidents on the job. Certainly more claims would result, since a controller could very nearly disqualify himself by alleging gradual stress. As a third alternative, suppose that to counter the potential fraud problem Congress were to drop FECA coverage for mental-mental claims, those for which determining authenticity is most difficult. Even if the second career program were still operating, this would discriminate against those legitimate victims of occupational stress simply because they didn't develop an ulcer or suffer heart failure.

None of these alternatives seems a desirable solution. Another solution, embedded in proposed amendments to FECA, appears more promising.[10] Some officials have suggested that compensation abuse is the law's fault, not the worker's, because it offers the temptation of generous benefits. Without endorsing this statement entirely, we suggest that it contains an element of truth supported by empirical study over the last decade. One provision of the proposed changes to FECA acts upon it. Under it, OWCP would reduce compensation payments to 80 percent of a proxy for the employee's spendable (after-tax) income. This eliminates the incentive of tax-free compensation payments exceeding normal take-home pay for high salary employees. By imposing some financial sacrifice, the provision would

---

10. The proposed amendments are in a bill entitled Federal Employees' Reemployment and Compensation Amendments of 1981. H.R. 4388, 97th Congress, 1st sess., August 4, 1981.

weed out some proportion of illegitimate claims. At the same time, it would reduce the need for job performance evidence as a test for claim authenticity.

Barring advances in medical diagnostic technology, mental disabilities will remain hard to assess. A commitment to covering this class of occupational disease requires some measures to preserve the integrity of the compensation program. It seems reasonable that if examiners can't tell whether abuse is occurring, steps should be taken to reduce the gains from abuse.

6 · EXPERIENCE-RATING AND INJURY PREVENTION

James R. Chelius and Robert S. Smith

The workers' compensation premiums paid by individual firms are based on an industrywide, or manual, rate. When a firm is sufficiently large to generate a manual-rate premium averaging $2,500 per year over a two-year period, its own injury experience is used to modify (experience-rate) the industrywide rate in setting the firm's premium. Experience-rated premiums are essentially a weighted combination of industrywide and firm-specific insurable losses, where the weight given to firm-specific losses rises with firm size, the industrywide risk of injury, and the level of workers' compensation benefits in the state. Although the criteria vary from state to state and industry to industry, generally only very large firms (a thousand or more employees, say) are self-rated; that is, only in large firms is full weight given to firm-specific losses with no weight given to industrywide losses. A decade ago, it was estimated that more than 80 percent of all employees worked for firms that were not fully self-rated (Russell 1973), and while this figure may have fallen some since then, it is clear that, in the typical case, workers' compensation premiums only partially reflect a firm's own injury experience.

Given that one of the major goals of the workers' compensation

The authors wish to thank John Inzana of the Bureau of Labor Statistics for his cooperation in assembling the data and are grateful to Richard Victor of the Rand Corporation for several helpful comments. Both are absolved of responsibility for any errors. —J.R.C. and R.S.S.

system is to encourage firms to adopt safer technologies and procedures, the lack of complete self-rating could pose a substantial problem. Suppose that a firm is thinking of spending $2,000 a year to maintain a device that will prevent injuries for which the annual compensable loss would total $2,500. If the firm were completely manual-rated, its premium savings attendant to this expenditure would be zero; that is, its premium reflects industrywide losses alone, and these losses are not appreciably affected by the $2,500 reduction in one firm. If the firm were partially experience-rated, premium savings would lie somewhere between zero and $2,500 — with a good chance that they might be below the $2,000 cost. Only if the firm were fully self-rated (or very close to it) would firms have an unequivocal incentive to undertake the hypothesized safety expenditure. A profit-maximizing firm would simply not spend $2,000 if the returns from that expenditure were less.[1]

Various remedies have been proposed to strengthen the safety incentives of workers' compensation. The National Commission on State Workmen's Compensation Laws (1973) advocated using a greater degree of experience rating, although it recognized the large administrative costs involved. James Chelius (1977) has suggested that workers be forced to bear more of their injury costs, but be given the right to sue their employers for gross-negligence — which would improve incentives for employees and employers to undertake precautions. Finally, Robert Smith (1981) has recently suggested that workers' compensation policies be written with a $500 deductible clause (that is, companies would have to self-insure against the first $500 of injury losses, which would prevent the spreading of losses to other parties for most injuries). A similar movement in the direction of self-insurance could be achieved by writing such policies with coinsurance clauses.

These suggestions have their merits and drawbacks. It is not our purpose here to evaluate these attributes; rather, we simply note that all are based on the *presumption* that the social losses associated with workers' compensation are large enough to merit attention. This

1. Firms may, of course, find that safety investments reduce injury-related costs (such as loss of worker morale) not covered by workers' compensation, but these noncompensable costs do not change the thrust of the argument here.

presumption, however, has not been empirically established. Russell (1973), for example, found evidence that large firms — which generally face the greatest degree of experience-rating — have lower injury rates than medium-sized firms. But very small firms generally also have lower injury rates than medium-sized firms, meaning that the causation between injuries and experience-rating is not at all clear. Moreover, Chelius (1982) found that workers in states with high benefits are more likely to be injured, which suggests that employees might have reduced incentives to take precautions as greater proportions of injury losses are compensated. While this finding is clearly relevant to an assessment of the safety incentives inherent in workers' compensation, it does not address the issue of how responsive *employers* are to differences in experience rating.

The present research is designed to gain some insight on the degree to which employer injury prevention activities are affected by the insurance arrangements used in workers' compensation. We attempt to solve some knotty empirical issues that have hindered a credible investigation of this issue, and our results suggest that the safety effects of experience-rating may be fairly small. If these results stand up in subsequent replications or refinements, one could argue that the safety incentive problem associated with the use of manual-rated workers' compensation insurance is small and perhaps not worth worrying about.

The Empirical Tests

Workers' compensation insurance premiums affect the incentives for a firm to adopt risk-reducing technologies or procedures only to the extent that these premiums are related to the firm's own injury experience. The weight given to the particular firm's experience rises with the firms' expected losses. Firms that are large, in risky industries, or in high-wage or high-benefit states are more likely to be experience-rated than firms that are not.[2] If a firm's premium is completely independent of its own experience, then efforts to reduce risk will bring no benefits in the form of reduced premiums. But the greater the weight

2. Experience-rated modifications to the manual rate are discussed in Russell (1973), pp. 33–36.

placed on the firm's own injury experience, the greater will be the premium reduction (other things equal) if efforts to reduce risk are successful. The actual premium reduction attendant to a given decline in the injury rate is a product of the weight *W* given to the firm's injury experience and the average benefits *B* for each injury avoided.

One problem in associating premium-related safety incentives with employer safety-related behavior is that the experience-rating formula is the same for each industry and state. This homogeneity robs any cross-state sample of some of its potential variance. Fortunately, however, wages and benefit formulas do vary across states (even within industries), so that expected premium savings *S* associated with a given decline in workplace accidents do vary in a cross-state sample. In particular, as *B*, average benefits, rises (*B* is affected by both wages and the generosity of benefit formulas), the marginal savings from a dollar spent on preventing injuries will rise — and this rise should induce more injury prevention activities on the part of employers.

We might expect, then, that experience-rated firms in states with high benefit levels will have lower injury rates than those of the same industry in states with smaller benefit levels *if* state safety laws, the technology in each industry, and incentives for worker precautions are similar across states. State safety regulations, to the extent states have their own safety programs at all, must be at least as stringent as those in force at the federal level under the Occupational Safety and Health Act (OSHA). Because the incentives to go beyond OSHA in stringency are small and because the effects of government safety programs have not been demonstrated, assuming essential homogeneity of state regulation is probably harmless.[3] Likewise, there is no reason to believe that production techniques *within* industries vary other than randomly (at least with respect to *B*) across states. But the finding reported earlier that higher workers' compensation benefits may induce worker carelessness is an obvious problem with which our research design must be concerned.

---

3. The only study to show any benign effects associated with OSHA at the firm level was Smith (1979b), and even these effects were confined to one of the years studied (1973). Studies covering the years after 1974 have found no significant effects. Moreover, an earlier study by Chelius (1973) found no benign effects associated with pre-OSHA state safety laws.

Worker behavior will be affected (if at all) by B; it will not be affected by the marginal premium cost (S). Thus, in a given industry and state, all workers — regardless of whether they work for an experience-rated firm or not — have about the same incentives to take precautions. Firms, however, will not. S will equal zero for some firms and WB for others (the experience-rated). Thus, injury rates in experience-rated firms should be lower relative to those of non-experience-rated firms in the same industry in states where workers' compensation benefits are larger. This prediction arises from the fact that, holding W constant, the difference between an S of zero and an S of WB rises with B.

The first step in testing this prediction is to define a variable F_{jk}, which equals the difference between the average injury rates of large, experience-rated firms (f^*_{jk}) and those of small, manual-rated firms (f_{jk}) in industry j and state k:

$$F_{jk} = f^*_{jk} - f_{jk}. \tag{6.1}$$

The prediction is that in states and industries where workers' compensation benefits (B_{jk}) are larger, f^*_{jk} will be smaller than f_{jk}, other things equal, because of the increased marginal premium savings among experience-rated firms relative to those that are not experience-rated.

Because we have assumed that within industries production technologies and government safety regulations will be either homogeneous or randomly different across states, empirical testing of our prediction concerning the effects of experience-rating is best done with a sample of data stratified by industry. Our approach was to estimate the following equation for each of fifteen two-digit manufacturing industries:

$$F_{jk} = a_{0j} + a_{1j} B_{jk} + e_j, j = 1 \ldots 15. \tag{6.2}$$

It is important that F_{jk} is to be calculated as the difference in injury rates between two groups of firms narrowly defined by size. Thus, the constant (a_{0j}) will capture any forces other than workers' compensation (for instance, differences in worker supervision, economies of scale in providing safety, worker specialization) that are creating differences in injury rates between these two size groups. Our prediction is, of course, that a_{1j} will be negative, reflecting that whatever inherent differences there are in injury rates across size groups in a particular industry, f^*_{jk} will be smaller relative to f_{jk} when B_{jk} is greater.

The Data

Neither of the two vectors of data needed to estimate equation 6.2 is readily available. Data on average injury rates by plant size category, industry, and state were available from the Bureau of Labor Statistics only for the year 1979 and only for the thirty-seven states participating in a cooperative, federal-state data gathering program. But for the year and states where injury rate data were available, we were able to obtain the number of cases involving lost workday cases per 100 workers, a measure of risk we have successfully used in past research. To reduce sampling variance in our injury rate data, only size-industry-state cells including 1,000,000 labor hours or more were included in our analyses.

Workers' compensation benefits by industry and state were calculated in three steps. First, we determined for each state the legally mandated workers' compensation replacement rate for temporary total disability. Second, we multiplied this replacement rate by the average weekly earnings in each two-digit industry within each state. Finally, we compared that estimated weekly benefit with the maximum benefit allowed in each state, and where the calculated rate exceeded the maximum we replaced the former with the latter in our vector of data.

This measure of benefits is crude in many ways. It does not take into account the waiting periods, typical number of weeks lost per injury, or size of awards for death or permanent disability. A prior study that carefully accounted for these factors, however, calculated benefit levels that were highly correlated (0.9) with a series estimated using the cruder techniques used here (Chelius 1973). Hence, for sake of thrift we elected to use the simpler methodology for creating estimates of B_{jk}.

Empirical Estimates

Four estimates of equation 6.2 were made for each of fifteen two-digit

TABLE 6.1
Estimates of the Coefficient of B_{jk} in Equation 6.2
(standard errors in parentheses)

Industry (Sample Size)	Estimate A	Estimate B	Estimate C	Estimate D
Food (32)	−.03(.03)	.02(.03)	.02(.03)	.07(.04)
Textiles (18)	.07(.05)	.04(.07)	−.01(.07)	−.03(.08)
Apparel (25)	−.06(.03)*	.02(.02)	−.10(.03)*	−.02(.03)
Lumber (19)	−.07(.04)	−.09(.03)*	−.01(.04)	−.03(.02)
Furniture (15)	.04(.04)	−.01(.05)	.02(.07)	−.03(.05)
Paper (22)	.01(.01)	−.02(.02)	.01(.02)	−.02(.03)
Printing (24)†	−.02(.02)	−.01(.02)	−.02(.02)	−.01(.02)
Chemical (24)	−.01(.02)	−.02(.02)	.02(.02)	.01(.02)
Rubber (13)	.06(.05)	−.02(.02)	.08(.05)	.00(.06)
Stone and clay (19)	.04(.03)	.08(.03)	.01(.04)	.05(.03)
Primary metals (20)	.04(.04)	.02(.04)	.06(.05)	.04(.04)
Fabricated metals (20)	−.01(.03)	.02(.02)	−.01(.03)	.01(.02)
Non-electrical machinery (20)	.04(.03)	−.01(.02)	.03(.04)	−.01(.03)
Electrical machinery (18)	−.05(.02)*	−.01(.02)	−.04(.02)*	−.00(.03)
Motor vehicle (16)	−.03(.02)	−.01(.03)	−.01(.03)	.01(.03)

Notes: *Indicates statistical significance at the .05 level using a one-tail test.
† Sample size = 25 for estimates C and D.

level manufacturing industries. Each of the four estimates involved a different comparison of plant size categories:

Estimate A group 6 (500–999 employees) compared with group 1 (1–19 employees)

Estimate B group 5 (250–499 employees) compared with group 1

Estimate C group 6 compared with group 2 (20–49 employees)

Estimate D group 5 compared with group 2

Plants in neither group 6 nor group 5 are large enough to guarantee that all plants in those groups are fully experience-rated; however, the scarcity of plants larger than these shrank sample sizes drastically, making it unfeasible to use them. Similarly, it is possible that some plants in groups 1 and 2 were not subject to complete manual rating (that is, they were partially experience-rated). Plants in groups 5 and 6, however, were much *closer* to full experience rating than those in groups 1 and 2, so these data do test our hypothesis.

Table 6.1 summarizes our estimates of the coefficient on B_{jk}. One can readily see that of the sixty estimates, only five are negative and significant at or above the 0.05 level. Further, only three industries are represented in these five estimates: apparel, lumber and wood products, and electrical machinery.

The small sample sizes within each industry may be responsible, of course, for the small number of statistically significant effects.[4] However, the large number of estimates made possible by our data set permit us to use a nonparametric technique, the "sign test," to test our hypothesis. With only thirty-two negative estimates out of sixty, however, we cannot reject the null hypothesis that the degree of experience-rating has no effect on employer behavior.

One problem with our empirical test is that workers' compensation premiums are set by *firm,* whereas our data on employment size are by *plant.* If the small plants in our sample are branches of larger firms, they may be subject to a greater degree of experience-rating than we assumed. There was no direct way to investigate this problem with the data from the Bureau of Labor Statistics. We were able, however, to use the 1976 Employer Expenditures for Employee Compensation (EEEC) data file to calculate, by industry, the percentage of small plants that were part of larger firms (at least in that data set). Of the 146 manufacturing plants with 1–49 employees, 24 (16 percent) were part of a larger entity, suggesting that our test may be flawed to a limited extent.

It is interesting to note, however, that there were four two-digit industries on the EEEC tape in which none of the small plants were branches of larger firms: lumber and wood products, printing and

4. We did *pool* all industry and state data to generate a sample of 305 observations. Estimating equation 6.2 with dummy variables for each industry did not yield results that were qualitatively different than those reported in table 6.1. But the implicit assumption that the coefficient B_{jk} is the same across all industries is clearly inappropriate, and therefore our pooled results are not convincing.

publishing, primary metals, and non-electrical machinery.[5] Of the sixteen estimates of B_{jk} in table 6.1 for these industries, ten are negative — still too few to achieve statistical significance using the sign test. In contrast, however, in the industries where branch plants were most commonly found among small establishments, only five of sixteen coefficients were negative.[6]

It appears, then, that experience-rating, or the lack of it, in workers' compensation has no observable effect on employer behavior. While the preponderance of evidence from this study certainly supports this conclusion, there is an inkling to the contrary: a slight majority of the sixty coefficients were negative. The results are more consistent with our theoretical expectations in those industries where the problem of branch plants seems smallest. The fact that these results, however weak, follow a pattern suggests that employer behavior may not be completely random with respect to experience-rating. Perhaps we are looking for the hypothesized effects with an instrument that is too blunt.

One way to sharpen the study is to use a more detailed, time-consuming procedure to calculate — for each state and industry — a much more precise estimate of the marginal premium costs. We have seen that marginal premium costs involve a weighting factor W multiplied by B_{jk}. We assumed that, within industries, W was more or less constant across states; however, more exact calculations of W could be made by simulating the rate-setting procedure in each state. Further, as acknowledged above, our measure of B_{jk} is crude and could be made more accurate by inclusion of permanent disability and death benefits. We intend to continue our investigation of this question by specifying the marginal premium cost in greater detail.

What can be said at this point is that if experience-rating has an effect on employer behavior, it is not large enough to observe using our simple proxy for marginal premium cost. That is, if effects of

---

5. Sample sizes were small in each case: fourteen firms in the lumber industry, nine in printing, eight in primary metals, and twenty in nonelectrical machinery. There were two other industries, rubber and transportation equipment, in which all small plants were separate entities, but those sample sizes were four and two respectively.

6. The industries represented were food, chemicals, stone and clay, and fabricated metals. All told, 16 of 55 (29 percent) small plants in these industries were branches of larger firms.

experience-rating are present, they will be observed only by using an independent variable with much less error in it.[7] This implies that the behavioral effects may indeed be so small that no major revisions in experience-rating are warranted.

7. Errors in an independent variable tend to bias the estimated coefficients toward zero.

7 · WORK DISINCENTIVES OF BENEFIT PAYMENTS

William G. Johnson

This is one of the first empirical studies of the work disincentives of workers' compensation payments. The results apply to workers whom the New York State Workers' Compensation Board found eligible for scheduled benefits for permanent partial disability.

Workers' compensation benefits are paid to compensate for the wage losses and medical care costs of work related illnesses or injury. The reasons suggested for less than complete compensation of wage losses are to provide incentives to all workers to avoid workplace injury and to encourage injured workers to return to work. In the first instance, the incentive operates by increasing the expected costs of injury to all workers and, in the second, by minimizing the income effect of benefits on labor supply.

Many states' permanent partial disability benefits are determined by a schedule that defines the amount and duration of the payments that are made for various types of injuries and degrees of severity. No reference is made to the worker's wage losses in the period during which benefits are paid. Although a relatively small proportion of workers' compensation beneficiaries receive payments of

I would like to thank Kathy O'Reilly, Cynthia Lowe, and Ginny Rapant for clerical assistance and Ed Heler, Nancy White, and Tod Porter for statistical work. All are members of the Health Studies Program at the Maxwell School, Syracuse University.

Support for the preparation of this study was provided by the National Council on Compensation Insurance and the Health Studies Program, the Maxwell School, Syracuse University. — W.G.J.

this type, the payments represent more than 60 percent of total benefits in most states.

Information on the magnitude of work disincentives under existing practices is especially important now because of the trend away from scheduled benefits to payments that vary with post-injury wage losses. The movement to wage-loss benefits is an attempt to provide more adequate benefits and, by relating benefits to actual losses, to distribute the benefits more equitably. Economic theory predicts that the disincentive effects of benefits defined by wage losses are likely to be greater than those under the current scheduled system. The policy question, then, is to what extent will increases in adequacy and equity increase costs by extending the duration of time between injury and return to work? An answer is possible only if one can compare disincentive effects under the two alternatives.

Economic studies of the disincentive effects of disability benefits, unemployment insurance, and public assistance are based on the following propositions: (1) individuals receive no benefit from work other than the wages and fringe benefits paid to them; (2) individuals improve their well-being by consuming goods and by consuming leisure time; (3) an increase in income, with no change in wages or prices of goods, will induce individuals to increase their consumption of both goods and leisure time; and (4) the time available for work or leisure equals total time (twenty-four hours) less the hours required for personal maintenance (e.g., sleeping, eating). Since time is fixed in amount, increases in leisure time can only be obtained by reducing hours spent at work, subject to the limits imposed by the time required for personal maintenance.

Embedded in these basic concepts is the idea that leisure time can be thought of as a commodity that can be purchased. Since leisure time can be increased only by reducing hours of work, the cost of an hour of leisure time is the wage income for that hour.

Most disability benefit programs and all welfare programs limit wages an individual can earn if he or she wishes to receive benefits. For these beneficiaries, the price of time equals the difference between the wage rate and the benefit rate. This reduction in the price of time will, the theory predicts, induce individuals to consume more leisure (money income is not increased, so the consumption of goods is expected to be unchanged). Since hours devoted to leisure can be increased only by reducing work hours, the full effect of benefits is a disincentive to work.

Scheduled workers' compensation benefits are one of the few disability benefits paid without reference to the wage income or hours of work of the beneficiary (veteran's compensation is another example). The income received from workers' compensation can be spent directly on goods and services or, indirectly, by giving up wage income, on leisure time. The theory predicts that the work disincentive will be smaller than in the case of wage-conditioned benefits, because the income from workers' compensation is divided between goods and leisure time, while wage-conditioned benefits can be used only to purchase leisure.

Another characteristic of scheduled permanent partial benefits reinforces this prediction: the payment of benefits is limited in duration, and the worker knows the duration at the time the award is made. The predictions of theory (Ghez and Becker 1975) and the results of social experiments imply that individuals respond to transitory changes in income differently than to permanent changes. In general, it is believed that the disincentive effects of benefit payments, in any given time period, are less if the recipient knows that the payment will terminate before the end of his or her working life. Since the duration of the benefits in the present study is rather short, one would predict that the disincentive effect of a given annual amount is small.

The implications of the theory, therefore, are that any increase in nonwage income has some disincentive effects on work: that the disincentive effects of a pure income transfer (i.e., scheduled benefits) are smaller than the disincentives associated with payments conditioned on the wage income of the recipient; and that payments of limited duration embody smaller work disincentives than those paid for the remaining working life of the individual.

These well-known hypotheses have been widely used to study the disincentives in unemployment insurance and negative income tax schemes. Their extension to workers' compensation benefits is straightforward, assuming that the effects of injury on productivity and on the time available for work can be adequately represented.

One of the most difficult problems in the evaluation of labor supply of workers whose health is impaired is the definition and measurement of health. The definitions used in this study are from the

World Health Organization's classification scheme (1980). An impairment is defined as " . . . any loss or abnormality of psychological, physiological, or anatomical structure or function." A disability is defined as " . . . any restriction or lack (resulting from an impairment) of ability to perform an activity in the manner or within the range considered normal for a human being."

Impairments may be temporary, as in the case of an acute illness, or permanent. The workers under study have permanent partial impairments; that is, their injuries produce a loss or abnormality that cannot be reduced through additional medical care.

Not all permanently impaired persons are work disabled. Whether or not an impairment leads to a restriction or lack of ability to work depends upon a number of factors, among them the physical requirements of a job and the way in which the required abilities are limited by the impairment. The loss of a hand, for example, may totally disable a laborer or partially disable a filing clerk but may not disable a computer programmer. Other factors known to influence the outcome of impairment are age (older persons are more likely to be disabled by a given impairment than younger persons); sex (the probability of work disability is higher for women than for men with the same impairment); education (the probability that an impairment is disabling varies inversely with education); and marital status (the probability of disability is less likely for married men and unmarried women than for others) (see Nagi 1969; Berkowitz and Johnson 1974; Lambrinos 1981; and Yelin, Nevitt, and Epstein 1980).

In general, as the severity of impairment increases so does the probability that the individual is work disabled. Impairments may be so severe that the individual is unable to work or so trivial that work capacity is unaffected. Since most injured workers are not at these extremes, however, the distinction is empirically important. Because of the many intervening factors, however, the prediction of disability based solely on the nature or severity of impairment is highly uncertain.

An impairment may limit productivity, hours of work, or both. I assume that productivity varies inversely with the severity of impairment. Certain impairments, regardless of the level of severity, also restrict the hours available for work. The productivity of a person who partially loses sight is severely impaired but time available for work is not affected. Someone who loses a limb may be limited in productivity, but may also require longer rest periods to maintain productivity. As rest or maintenance time is increased, the time available for both work and nonwork activities is reduced.

The Empirical Model

My empirical estimates are based on techniques that permit the identification of the independent effects of the different influences (e.g., impairment, wages, benefits) on the worker's decision concerning the number of hours that he or she will work.[1] The model consists of three equations: (1) a wage equation to estimate the offer wage for workers, whether or not they are employed; (2) an equation to predict the probability of labor force participation; and (3) an equation to analyze the determinants of variations in the hours of work of labor force participants.

The Wage Equation. To evaluate disincentives, one must estimate the wages that workers could earn if they worked. The post-injury wage differs from the pre-injury wage by the loss of productivity caused by the impairment. The loss can sometimes be offset by additional training or experience. Money wages may also increase due to the effects of price inflation.

Variations in our data make it necessary to use three different methods of estimation. As an estimate of the offer wage in 1971, we use the average weekly wage that each worker earned during 1970, twelve months before the accident. All the workers were employed at the time of injury, and the average weekly wage in 1970 is an excellent proxy for the wage workers could have earned in 1971. A measure of impairment, *TBI*, is used to represent the reduction in productivity that may occur due to the impairment.

An earnings function is used to estimate offer wages in 1974. This is the only year in which it is possible to estimate both a participation and an hours worked equation. The earnings function can be written as

---

1. I have adopted the usual model of labor supply in which utility is a function of market goods and leisure or nonmarket time. Nonmarket time is assumed normal and the prices and relative proportions of market goods are assumed to be fixed. I will be happy to supply further details on this and other models presented here.

$$lnW\mathrm{i} = a_0 + a_1ED + a_2EXP1 - a_3EXP1^2 + a_4EXP2 - a_5TBI + a_6RACE + a_7SEX + a_8UN + a_9\,Lambda \quad (7.1)$$

where: lnW = the natural log of the wage rate

$EXP1$ = general experience in the labor force (that is, age-education-6-$EXP2$)

$EXP2$ = firm-specific experience

TBI = a measure of the severity of impairment

$RACE$ = 1 if the person is white

Sex (SEX equals 1 if person is male) and union membership (*UN* equals 1 if a union member) are included as controls for the possible existence of wage differentials due to prejudice (+) or union bargaining power (+).

Lambda is Heckman's correction for selectivity bias (Heckman 1979).

Labor Force Participation. Approximately one-fifth of the workers left the labor force following their injury and were still out of the labor force in 1974 and 1975. If there are disincentive effects of workers' compensation, they should be observed in the estimates obtained from the participation equations as well as for the variations in hours of work among employed workers.

The workers may be out of the labor force in the post-injury period because: (1) maintenance time is at a level where the marginal utility of time exceeds the offer wage; (2) income in the form of workers' compensation benefits and increased consumption of leisure time is sufficient to overcome the loss of income occasioned by withdrawal from the labor force; or (3) because of the loss of productivity, the offer wage is less than the marginal utility of time.

The labor force participation equations in 1974 and 1975 are estimated using determinants of wages (the independent variables in equation 7.1) as the proxy for wages. The labor force participation equation can be written as

$$LFP = b_0 + b_1W - b_2WC - b_3M + b_4MRSEX - b_5RACE \quad (7.2)$$

where: *LFP* is a binary measure of participation; *WC* equals workers' compensation benefits received in the period in question; M is a measure of the time required for maintenance activities and is defined as the sum of the values reported for limitations in the worker's ability to sit or stand for extended periods of time; $MRSEX$ equals 1 if worker is a married man or an unmarried woman.

These estimates are not adjusted for taxes. The bias should be small, however, since the wage data do not include fringe benefits. For most of the workers in the sample, it is likely that the fringe benefits, which typically average from 17 to 18 percent of total compensation, and tax effects are nearly equal and opposite. In 1973 the average federal tax rate on incomes between $5,000 and $15,000 was approximately 17.5 percent (U.S., Bureau of the Census 1979, pp. 266).

Hours Worked Equation. The hours worked equation relates variations in hours worked among labor force participants to wage rates, workers' compensation benefits, constraints on time, and preferences. The relationship can be written as

$$Q = g_0 + g_1W - g_2WC - g_3M + g_5MRSEX. \tag{7.3}$$

There is one complicating factor: some of the workers also received social security disability insurance benefits. To be eligible for social security disability benefits the worker must be medically determined to be "permanently and totally disabled" and cannot earn more than an amount defined as "substantial gainful activity." The limit is less than the wage income of full-time work at the federal minimum wage rate. Too few persons received these social security benefits to estimate an expected benefit for each worker; most of the awards were received after 1971. To approximate the effect of social security disability benefits, the model is estimated with beneficiaries included then with beneficiaries excluded, and the results are compared. The effect of our exclusion is to assume that persons who subsequently received social security disability benefits were, as a group, facing a different set of incentives in 1971 than the other workers. This assumption seems reasonable since workers must be out of the labor force for six months before they can apply for social security disability benefits.

TABLE 7.1
Characteristics of Workers Receiving Scheduled Permanent Benefits, New York

In labor force, 1971	77%
In labor force, 1974	84%
In labor force, 1975	77%
Male	88%
White	87%
Union member	58%
Severity of impairment, mean	16.6%
Years of education, mean	10.1
Years of firm-specific experience, mean	6.9
Workers' compensation received, 1971, mean	$2,395
Workers' compensation received, 1974, mean	$ 843
Workers' compensation received, 1975, mean	$ 256

N = 432

The Data

Our data set is a representative sample of workers in New York State who received workers' compensation benefits for permanent partial or permanent total disability as a result of injuries incurred in 1970.

The data include the results of an hour-long interview of each worker conducted in 1975 and the summary case record maintained by the New York State Workers' Compensation Board (Makarushka et al. 1977). The sample includes only those workers whose impairment was greater than 9 percent of total bodily capacity as defined by the American Medical Association (1971). Since our interest is in labor force behavior, persons sixty-five years of age or older in 1975 are excluded. The data were further restricted to exclude social security disability beneficiaries and persons who received permanent total disability benefits. The characteristics of the 432 workers in the final data set are described in table 7.1.

The injury marks the beginning of a period in which temporary influences on labor supply may dominate worker behavior. One influence is the need to allocate time to medical care. Another influence results from the fact that workers' compensation agencies often consider evidence of wage loss in determining benefit amounts even though, once awarded, the benefits are not income tested. There are, therefore, incentives for workers to maximize wage losses during this period by remaining out of the labor force. These incentives are likely to be reinforced in cases where workers are represented by attorneys.

Once a course of acute medical care is completed and a benefit is determined, disincentive effects of workers' compensation can be evaluated as an independent influence, free of the confusion introduced by the temporary incentives. Our data permit analysis of the labor supply of injured workers at three points, 1971, 1974, 1975. Cheit (1962) concludes that injured workers complete all adjustments to the injury within five years. If this assumption is correct, the labor force behavior of the workers in 1975 represents the outcome of all adjustments to the injury and impairment. Unless changes not related to the injury intervene, one would expect the 1975 labor force participation rates to represent the future behavior of these workers.

A worker who is eligible for benefits may receive a lump sum settlement, a series of equal payments for a specified duration or a series of payments and a retroactive lump sum equal to payments owed for the period between the date of injury and the date of award. Sixty-three workers received a series of benefit payments. An additional 236 workers received a series of payments and a retroactive lump sum. Single lump sum settlements were received by the remaining 133 workers.

Single lump sum settlements provide the recipient with a number of options concerning the time distribution of consumption of the benefit. The pattern chosen will be determined by each individual's preferences between current and future consumption and market interest rates. Strictly speaking, the worker's options exist for his or her lifetime, but since the average lump sum is small, it would be unrealistic to allocate these sums over the lifetime since such an approach yields trivial amounts of imputed income in any year. Following Cheit's rationale (1962), I allocated the lump sum payments over five years from the time of the accident. Interest income that might be earned by investing the lump sum is not included. The estimated coefficients are the same as those that would be obtained by the inclusion of interest, since the data are from cross sections and the duration under study is the same for all the workers.

The duration of payments to workers who received some or all of their benefits in a series was less than one year for 77 percent of the cases. Approximately 53 percent of all cases were closed by the end of 1971; an additional 34.3 percent were closed in 1972, and all cases had been closed by the end of 1973. This indicates that the maximum disincentive effects of workers' compensation payments should be observed in 1971 and 1972. My data do not include 1972, so I focus on the disincentive effects in 1971.

Empirical Results

Labor Force Participation. Labor force participation equations were estimated, using probit, for 1971 and 1974. Significant disincentive effects are observed in 1971 (table 7.2). The elasticity of labor force participation relative to benefits is estimated to be −0.088. Evaluated at the mean value of labor force participation rates in 1971, the elasticity indicates that if benefits were 10 percent greater, the labor force participation rate would be 0.7700 versus 0.7769. The difference in participation rates translates into a reduction in labor force participation equal to approximately 2.8 workers. In other words, for each 10 percent increase in the average workers' compensation benefit, the labor supply among the 432 injured workers is reduced by approximately three worker years. Since we have not distinguished between those workers who were full-time and those who were part-time before they were injured, the term worker year should not be construed to necessarily mean year-round, full-time work.

The most curious aspect of the workers' labor force behavior is that a substantial portion of those who are injured leave the labor force and remain out of the labor force after benefit payments have ended. The highest participation rate is the 85 percent figure for the year 1974. For 1975, where the measure is based on a survey week response, the rate is the same as in 1971 (77 percent). In other words, most of those who return to the labor force do so within one year of

TABLE 7.2
Labor Force Participation,
Probit Estimates,
1971

Variable	ML Estimate of Parameter	Asymptotic *t*-ratio
CONSTANT	0.221	0.74
WAGE	0.004**	1.81
MRSEX	0.427*	1.97
*WC*71 ($100)	−0.009*	−2.82
TBI	−0.017*	−1.91
RACE	0.384**	1.85

Notes: Maintenance time is based on the 1975 interview and is not an appropriate measure of maintenance time in 1971.

The results for 1974 and 1975, which are described in tables 7.3 and 7.4, indicate that there are no significant disincentive effects of workers' compensation benefits in either year.

* significant at 0.05 level or better

** significant at the 0.10 level

$N = 373$

Log of likelihood = −18.60

Likelihood ratio chi-squared statistic for overall test of regression (−2.0 × log of likelihood ratio) = 37.206

Prob of chi-square > X with 5 D.F. = 0.000001

being injured. The labor force participation rates in 1974 and 1975 may reflect the fact that some workers, presumably those whose initial absence from work is of longer than average duration, return but do not remain in the labor force for very long (for a more complete discussion of this behavior, see Johnson, Cullinan and Curington 1979).

Because of the shortage of empirical studies of the disincentive effects of workers' compensation, there is no standard by which to judge the reality of these results. There are, however, some interesting similarities between my findings and Paul Fenn's study of ill and injured workers in Great Britain (1981). His data and method of estimation differ from mine in several respects, so only an approximate comparison is possible.

Fenn's estimate of the elasticity of the probability of return to work relative to the ratio of benefits to wages is −0.07. Both my estimate and Fenn's are substantially less than those obtained in studies of unemployment insurance; both are negative; and although his

TABLE 7.3
Labor Force Participation,
Probit Estimates,
1974

Variable	ML Estimate of Parameter	Asymptotic *t*-ratio
Constant	0.954**	1.83
General experience	−0.046**	−1.76
MRSEX	0.542*	2.61
WC74 ($100)	−0.016	−1.38
TBI	0.000	0.08
RACE	0.215	0.99
Specific experience	0.003	−0.25
Education	0.017	0.62
General experience squared	0.001	1.04
Maintenance time	−0.107**	−1.71
Union member	0.034	0.21

Notes: * significant at 0.05 level
** significant at 0.10 level
$N = 432$
Log of likelihood = −14.9881
Likelihood ratio chi-squared statistic for overall test of regression (−2.0 × log of likelihood ratio) = 29.976
Prob of chi-square > X with 10 *D.F.* = 0.000864

estimate applies to the replacement ratio, both are similar in magnitude. Another parallel that emerges from the two studies is the time distribution of the injured workers return to the labor force. Most of the workers who return to work are likely to do so within the first year. Fenn, using more disaggregated data, finds that the probability of return to work increases to a maximum in the first six months and then declines continuously to an asymptotic minimum in ten to eleven months. These comparisons, although limited by the differences between the studies, provide some support for the validity of our estimates.

One must be cautious, however, in interpreting this result. There are influences that cannot be adequately controlled that may be important determinants of labor force participation. This result could occur, for example, if workers who are negotiating for settle-

TABLE 7.4
Labor Force Participation
Probit Estimates,
1975

Variable	ML Estimate	Asymptotic *t*-ratio
Constant	0.939*	1.95
General experience	−0.035	−1.53
MRSEX	0.393*	1.95
*WC*75 ($100)	−0.162	−0.25
RACE	0.395*	1.97
Specific experience	0.012	0.95
Education	0.003	0.12
General experience squared	0.000	0.55
Maintenance time	−0.129*	−2.22
Union member	−0.199	−1.34

Notes: * significant at 0.05 level
Log of Likelihood Ratio = −21.0265
Chi-squared statistic for overall test of regression (−2.0 × log of likelihood ratio) = 42.053
Prob of chi-square > X with 10 D.F. = 0.000007

ments remain out of the labor force for a longer period than other equally impaired workers. One could argue that the result is a disincentive effect, but if it exists, it is a different kind than the effect that I attempt to estimate.

When the workers who received social security disability benefits are included, the results are essentially unchanged. The finding that workers' compensation benefits are a significant influence on participation in 1971 is not reversed. The estimated effect of benefits is larger, reflecting the omitted influence of the dual receipt of benefits by some workers. The influence of social security disability benefits appears to be important, therefore, as one which reinforces the income effect of workers' compensation but which influences only those workers whose impairments are well above the average level of severity.

Hours Worked. An equation of the general form of equation 7.3 was estimated for hours worked, using generalized least squares (GLS) for 1974. The results are described in table 7.5. No statistically signifi-

TABLE 7.5
Hours Worked,
1974

Variable	Parameter Estimate	*t* for *HO*: Parameter = 0
Intercept	24.029*	5.09
WAGE74	0.468	0.48
WC74 ($100)	−0.161	−0.99
Maintenance	−1.986*	−2.44
MRSEX	7.506	2.21

Notes: Adj. $R^2 = 0.0256$

WAGE74 = 4.013823 + .359003 *EDUCR* − .043515 *ESPG* − .000313294 *EXPGSQ* + .057136 *JOBTENUR* + .135899 *INTWRACE* + 1.111038 *UNIONMB* − 2.859626 *LAMBDA74* + 2.567439 *MRSEX*

* significant at 0.05 level or better

cant disincentive effect of workers' compensation benefits is found. The result is not very surprising, since few employed persons received cash benefits in 1974, thereby making any potential disincentive effect largely dependent on the imputed incomes received by workers whose settlements took the form of a single lump sum award in a prior year.

The fact that the *lambda* variable in the earnings function is not significant implies that there is no selectivity bias in the model.

Conclusions

The results indicate that workers' compensation benefits provide small but statistically significant work disincentives to permanently impaired workers in the year after their injury. The measured disincentive effect is essentially zero in the fourth and fifth year after the accident. These data do not reveal whether the disincentive effect persists into the second and third year after the injury.

Participation rates in 1974 and 1975 do not appear to be attributable to disincentive effects of permanent partial benefits. Is part of the explanation that persons leave the labor force because their impairments are so severe that they are simply unable to work? One would expect cases of this type to be confined to workers who have been defined to be permanently and totally disabled by the worker's compensation board or, if not so defined, have qualified for social security benefits. These cases are excluded from the data, and the coefficients obtained for the severity and maintenance time associated with physical impairments clearly demonstrate that little variance in participation stems from those influences. So it is extremely unlikely that nonparticipation among the remaining workers occurs as a result of physical limitations.

The labor force participation equations explain a rather low proportion of the variance in participation, although factors such as wages, severity of impairment, marital status, and sex are found to be important. Furthermore, descriptive data on the worker cohort suggest that rehabilitation services do not account for the difference nor does it imply that changes in the labor supply of other household members could make a significant difference (Makarushka and Johnson 1976). Considering these facts, one must conclude that a set of influences omitted from the models are important determinants of the labor supply of injured workers.

The model does not reflect employer reactions to the workers' injuries. It is likely that, in addition to the effect of impairment on the offer wage, the time required for recuperation provides employers with incentives to replace the injured worker, especially if employers fear that the worker's impairment increases the probability that the worker would be totally disabled by any subsequent injury. In such cases, the employer expects costs of compensation for impaired workers to be greater than for unimpaired workers. It is possible, therefore, that part of the variance in participation rates occurs because employers are less willing to hire impaired workers than unimpaired workers.

Notwithstanding conjectures about possible causes of the observed reduction in labor supply other than workers' compensation benefits, the only firm conclusion allowed by these results is that the disincentive effects of permanent partial scheduled benefits are small and limited in duration. It is also clear that the combination of social security disability benefits and workers' compensation is an influence on some permanent partial beneficiaries; the question of an overlap of workers' compensation and social security disability insurance is not limited to persons found eligible for permanent total disability benefits from workers' compensation.

A more fundamental result of the study is the implication that proposals to substitute wage-loss benefits for scheduled benefits are likely to increase disincentives to return to work. One of the dangers of the wage-loss approach is that it might increase the disincentive effects of workers' compensation benefits for permanent partial disability to a level approaching that observed for unemployment insurance. There might be perhaps as much as a sevenfold increase in the number of workers who do not return to work during the first year after an injury. The results here are not sufficiently general to suggest that this result would occur in every wage-loss system. It is an example, however, of the worst possible outcome of the shift toward wage-loss systems.

On the other hand, the wage-loss approach may be a badly needed remedy for the compensation of the large group of workers who never return to the labor force. The reasons for their behavior, although unknown, do not include a disincentives effect or the measurable (from these data) aspects of their impairments. If these workers are prevented from returning to work by a set of events triggered by the injury, then there is a rationale for better compensating them for wage losses. The net effect of wage-loss benefits appears to rest upon comparisons of the changes in the adequacy of compensation, adjusted for increases in costs that might occur through an increase in disincentives to return to work. The information presented here offers a starting point for such an investigation.

8 · THE INCENTIVE TO PREVENT INJURIES

James R. Chelius

The 1970s witnessed substantial changes in workers' compensation laws, which resulted in sizable benefit increases being available to most injured workers. While it is straightforward to conclude that the income security goal of workers' compensation has been enhanced by these changes, it is not obvious what impact they have had on the other major goal of the workers' compensation system: injury prevention. It is the purpose of this analysis to examine how recent changes in workers' compensation laws have influenced the allocation of resources to injury prevention. The empirical analysis indicates there is a substantial conflict between the income security and injury prevention goals of the workers' compensation system. If the joint achievement of these goals is to be optimized, it is necessary that the nature of this conflict be recognized.

An increase in workers' compensation benefit levels represents an increase in the cost of an injury to the employer and a decrease in cost to the employee. Economic theory predicts that when the cost of something increases the quantity decreases, and when the cost of something decreases the quantity increases. Therefore, increases in workers' compensation benefit levels may have two opposite effects on the level of resources devoted to prevention, a greater quantity of prevention demanded by employers and a lesser quantity demanded by employees. Since no theoretical analysis can predict which force is stronger, it is the role of the empirical analysis to determine the net result of the change.

We frequently sanctify injury prevention in a manner that implies prevention at any cost. Our actual behavior, however, appears to be based on the notion that injury prevention is the result of an implicit or explicit weighing of the advantages and disadvantages of prevention: the amount of resources devoted to preventing injuries depends on the cost of preventing an injury compared to the benefit of avoiding an injury. The probabilistic nature of danger makes the calculation difficult, but does not change its essence. The level of prevention may thus be manipulated by changing either the costs or the rewards of prevention. Since the workers' compensation system does not influence the cost of prevention, the present analysis will be confined to possible rewards of reduced injuries. To do this, we must consider the advantages of prevention to both employers and employees, both of whom are possible beneficiaries of the avoided injuries.

In physics, it is insightful to analyze the behavior of forces as they would operate in a vacuum. Similarly, analysis of the operation of economic forces in a frictionless environment is a useful first step. In such an environment wages would perfectly reflect the burden on the employee resulting from dangerous work. Other things being equal, a job with greater danger would have commensurately greater pay. This results not from the employer's largess, but from the necessity of paying higher wages to attract workers to the dangerous jobs. Higher workers' compensation benefits, by reducing the financial burden of being injured, would commensurately reduce the wage premium needed to attract workers to dangerous jobs. But while the reduced wages to employees represent a lightened burden to employers, the employer is now liable for greater workers' compensation benefits. Thus, increased workers' compensation benefits simply cause wage premiums to be reduced, with no change in the total reward for prevention to either the employee or the employer.

While the labor market may not operate as smoothly as envisioned by this paradigm, there is a substantial body of evidence indicating wage premiums are paid for hazardous work (Smith 1979a). The presence of such premiums, however, does not mean that workers' compensation premiums are irrelevant to the allocation of resources to prevention. The frictions likely to exist in any labor market make it possible that workers' compensation benefit levels change the rewards for prevention and, therefore, the level of resources devoted to such activity. These frictions might include factors such as the lack of information about riskiness in various jobs and the difficulty of enforcing agreements between employers and employees as to the appropriate level of prevention. The presence of such frictions makes a relationship between workers' compensation benefit levels and prevention activities plausible.

While wage premiums are implicit and subtle, workers' compensation benefits are explicit and obvious. Higher benefits that do not result in correspondingly lower wages would increase the costs of worker injuries to the employer and decrease the cost to the employee. If higher benefits do not have the same effect on employers and employees, the net effect of higher benefits would be to change the total prevention effort. Whether or not such an effect exists is a question for the present analysis.

The Empirical Test

Occupational injury rates are the best available measure of the resources devoted to prevention, and excellent injury rate data are available for many states over the recent period of changes in workers' compensation laws. Twenty-eight states have data for the years 1972 through 1978 for manufacturing industries.[1] In addition to workers' compensation benefits, other factors might influence injury rates. These include industrial characteristics, such as technology and plant size, and time-dependent factors, such as the business cycle and labor force characteristics. In order to correct for all these influences, the injury rate variable was expressed as the ratio of the injury rate for a specific industry, state, and year to the injury rate for that industry throughout the United States in that year; the injury rate variable represents the value of the specific observation relative to the more general industry and year. Thus, the effects of an inherently dangerous (or safe) technology would be reflected in both the numerator and denominator. Similarly, by making each ratio specific to a year, the

1. John Inzana of the U.S. Department of Labor, Bureau of Labor Statistics, was very helpful in assembling these data. These state and two-digit SIC industry-specific data were collected as part of the OSHA data collection program. Aggregates of these data are reported in publications such as U.S., Department of Labor, Bureau of Labor Statistics (1980c).

characteristics of that period would be felt in both the specific state and the whole country.

Workers' compensation benefits were measured relative to wages for each state, industry, and year observation. This puts such benefits in the form most relevant to employers and employees and corrects for the changing value of the dollar over this period. The resulting variable is labeled income replacement.[2] In addition to the rate at which benefits replace income, the period for which employees must wait for benefits is also potentially relevant. Therefore, a variable representing the number of days in the waiting period was added. In order to formulate the benefit and waiting period variables in a manner parallel to the injury rate variable, they were structured as the ratio of income replacement and waiting period for the particular observation relative to the value of those variables for the industry and year aggregate of those variables. This specification implies the following equation as the basis for estimating the empirical relationship.

$$\frac{\text{injury rate}_{itj}}{\text{injury rate}_{itus}} = \alpha_1 + \alpha_2 \frac{\text{income replacement}_{itj}}{\text{income replacement}_{itus}} + \alpha_3 \frac{\text{wait}_{itj}}{\text{wait}_{itus}} + \epsilon$$

where: injury rate $_{itj}$ = the number of injuries per 100 full-time workers in industry i, year t, and state j

injury rate $_{itus}$ = the average number of injuries per 100 full-time workers in industry i, year t, for the United States

income replacement $_{itj}$ = the ratio of weekly workers' compensation benefits to weekly wages in industry i, year t, and state j

income replacement $_{itus}$ = the ratio of weekly workers' compensation benefits to weekly wages for industry i, time t, for the United States

2. Benefits were measured for temporary total injuries, the most common type. Benefits for temporary total injuries are highly correlated with the level of benefits for other types of injuries.

$wait_{itj}$	= the length of the waiting period for benefits in industry i, year t, and state j
$wait_{itus}$	= the length of the waiting period for benefits in industry i, year t, for the United States
ϵ	= error term.

Overall injury losses are best approximated by the total number of lost workdays per hundred full-time employees. The frequency and severity (days per case) rates of these injuries are also interesting. The employee response to workers' compensation benefits is most likely to be visible in the frequency rate. Since pain and suffering continue as deterrents for more serious injuries, regardless of benefit levels, any influence on employees is more likely to be reflected in the frequency rate. The relationship between injuries and income replacement levels was estimated using regression analysis. The equa-

TABLE 8.1
Injury Rates and
Income Replacement

Dependent Variable	Constant	Income Replacement Coefficient (standard error)	Waiting Period (standard error)
Lost workdays	1.02	0.16 (0.05)*	−0.02 (0.55)
Frequency rate	0.97	0.22 (0.05)*	−0.04 (0.04)
Days per case	1.03	−0.05 (0.03)	0.01 (0.03)

Sources: The Chamber of Commerce of the United States (1972–75) and the *Report of the Full Compliance Subcommittee of the Ad Hoc Committee to Consider State Compliance with Workers' Compensation Recommended Standards* (1976). Wage data used as a part of the benefit calculation are from U.S. Department of Labor, Bureau of Labor Statistics (1979a). Laws were evaluated as of January 1 of each year. Injury data are from the unpublished files of the Bureau of Labor Statistics.

Notes: Occupational injuries are defined as any injury (such as a cut, fracture, sprain, or amputation) that results from a work accident or from exposure in the work environment. Occupational illnesses are excluded. In order to ensure comparable wage data, only manufacturing industries are included. The equations were estimated using dummy variables for each state except California to allow for non-workers' compensation differences across states.

$n = 1967$

* Significant at the 0.01 level

tion was estimated using three versions of the dependent variable: the total lost workday rate, frequency rate of lost workday cases, and the severity rate (average number of days per case). The results are presented in table 8.1.

The positive and significant coefficient for income replacement indicates that, on average, higher levels of income replacement are associated with more lost workdays. A higher frequency rate is also associated with higher levels of income replacement. There appears to be no significant relationship between the average number of days per case and the level of income replacement.[3] There was no significant relationship between the length of the waiting period and any of the injury rates.

The Policy Implications

The positive relationship between income replacement and both the total number of lost workdays rate and the frequency rate indicates that workers' compensation influences more than just the income security of injured workers. Unfortunately, the well-being of workers who are *already* injured is usually the only criterion used in policy decisions about benefit levels. While it is obviously a value judgment as to how much weight to place on the prevention goal as opposed to the income security goal of workers' compensation, it is important to recognize that there is a conflict between them.

In order to make this conflict between prevention and security more vivid, a comparison of injury rates is presented. Injury rates within categories for state, industry, and year that conformed to the benefit recommendations of the National Commission on State Workmen's Compensation Laws were compared to observations within categories that did not meet these recommendations. The

3. Previous work using 1972–75 data found an insignificant relationship between income replacement and the total lost workdays rate. This smaller data set yielded a positive and significant relationship between frequency and income replacement and a negative and significant relationship between days per case and income replacement (Chelius 1982). Cross-sectional work using only 1967 frequency rates also found a positive and significant relationship between benefit levels and injury rates (Chelius 1977). The lack of evidence of a relationship between the average number of days per case and the level of income may be due to an increase in short-duration cases induced by benefit changes.

TABLE 8.2
A Comparison of Injury Rates

	Observations Meeting the Proposed Standards	Observations Not Meeting the Proposed Standards	Absolute *t*-value of Differences
Lost workdays	1.15	.97	7.17*
Frequency rate	1.15	.97	8.13*
Days per case	0.99	1.00	.55
Sample size	327	1,640	

Sources: The Chamber of Commerce of the United States (1972–75) and the *Report of the Full Compliance Subcommittee of the Ad Hoc Committee to Consider State Compliance with Workers' Compensation Recommended Standards* (1976). Wage data used as a part of the benefit calculation are from U.S. Department of Labor, Bureau of Labor Statistics (1979a). Laws were evaluated as of January 1 of each year. Injury data are from the unpublished files of the Bureau of Labor Statistics.

Notes: Occupational injuries are defined as any injury (such as a cut, fracture, sprain, or amputation) that results from a work accident or from exposure in the work environment. Occupational illnesses are excluded. In order to ensure comparable wage data, only manufacturing industries are included. The equations were estimated using dummy variables for each state except California to allow for non-workers' compensation differences across states.

$n = 1967$

* Significant at the 0.01 level

commission recommended benefits be 66.66 percent of an injured worker's wages up to a maximum of 100 percent of the state's average weekly wage. A three-day waiting period was also suggested. This standard was chosen as the basis for comparison simply because most debates over appropriate benefit levels refer to the commission's work and its recommendations. Table 8.2 summarizes the injury experiences of those observations within industry, state, year categories meeting these two basic recommendations and those that do not. The total number of lost workdays and the frequency rate were significantly higher in the environments meeting the commission's proposed standards.

The conclusion that higher benefit levels are associated with higher injury rates does not demonstrate the overall superiority of a lower benefit policy. It is, more simply, an empirical finding that there is a conflict between income security and prevention. The balanced achievement of both goals can therefore be enhanced if attention is paid to this trade-off between security and prevention.

9 · DEMOGRAPHIC AND ECONOMIC CHANGE AND THE COSTS OF WORKERS' COMPENSATION

Alan E. Dillingham

The extent of change, both past and future, in the demographic composition of the work force is well documented, but virtually nothing is known about the effect of these changes on workplace injuries and workers' compensation costs. The purpose of this study is to determine the extent to which changes in the demographic characteristics of the work force and the industry-occupation mix of employment have an effect on workers' compensation costs. While the quantitative effect of demographic change on costs can be only crudely estimated at this time, it is important to both policymakers and the insurance industry that these less obvious influences on workers' compensation costs be analyzed.

Compensation cost per million employee hours of exposure is estimated with a linear model constructed from a unique set of employment and workers' compensation system data for one state. The model is then used to account for the changes in compensation cost per million employee hours between 1960 and 1980 and to predict future changes in compensation cost.

This research was partially funded by the Workers' Compensation Project of the University of Connecticut and the Employment Standards Administration, U.S. Department of Labor. — A.E.D.

The Calculation of Costs

Workers' compensation costs are determined by both the number of injuries and the cost of compensation per injury, which includes the fractional replacement of lost wages, medical expenses, and disability or death awards. The compensation cost for each injury is determined by the benefit provisions of the law, the worker's wage, the severity of the injury, and the worker's personal characteristics (e.g., state of health at the time of the accident). The total compensation cost is the sum of compensation costs for every compensated injury.

The effect of the provisions of the law, and changes therein, on cost per injury has received considerable attention in recent years because benefit reform and increased overall workers' compensation costs have occurred simultaneously. But the effect of variation in the annual number of injuries on total costs has until recently received little attention. Variation in the annual number of injuries might occur for at least two reasons. Liberalized benefits may alter worker behavior such that higher injury frequency rates arise (Butler and Worrall 1981). Second, broad changes in the characteristics of the economy or its labor force may alter the number of workplace injuries. These changes are the focus of this study.

Suppose that both the rate of injury (i.e., the number of injuries per five hundred full-time workers) and the number of employed full-time workers for an industry is known; their product yields the number of expected injuries in the industry. If the average compensation cost per injury can be calculated for the industry, it is possible to derive total compensation cost for the industry as the product of average cost and number of injuries — and economywide compensation cost is just the sum of industry costs across all industries.

With these group-specific data, a measure of unit cost can also be calculated. The product of the number of injuries per five hundred workers and the average cost of injuries for the group yields the compensation cost per five hundred workers, or per unit of exposure. Economywide compensation cost per unit of exposure can be expressed as the weighted sum of these group-specific unit costs. The weight given to each group is its relative employment share in the economy. Consequently, changes in economywide unit compensation cost over time can be caused by changes in group-specific injury rates, average compensation per injury, or both and by changes in the distribution of employment across groups and thus in the weighting of each group. Of course, total compensation cost will tend to rise over time as total employment grows.

This study derives injury rates and average compensation cost per injury for workers grouped by age, sex, industry and occupation. It then estimates changes in economywide per-unit compensation costs between 1960 and 1980 from the change in the age, sex, industry and occupation distribution of employment over this twenty-year period.

Injury Frequency and the Distribution of Employment

Systematic variation in injury frequency by broad firm and worker groups has been well documented. The frequency of injury varies by the age and sex of workers and the industry and occupation in which they are employed. If there exists much variation in injury frequency across these four dimensions of the labor market, then in a twenty-year period compensation costs could be markedly affected.

Previous studies have shown that younger workers have higher injury frequency rates than older workers (Dillingham 1981a; Kossoris 1940). In fact, OSHA critics have argued that one of the explanations for the increase in injury rates in the 1960s was the growing share of the labor force accounted for by young workers (see Chelius 1979 and Smith 1974). Furthermore, older workers tend to have more serious injuries, and to lose more work time per injury than younger workers (Dillingham 1979). Employers have expressed the belief that the injury experience of older workers is somehow more costly than that of other worker groups (Dillingham 1981a; Kossoris 1940).

Recent changes in the labor force age distribution have been marked. In 1960, 16.6 percent of the labor force was in the eighteen- to twenty-four-year-old category, while some 39 percent was over the age of forty-four. By 1979 the sixteen- to twenty-four-year-old group constituted over 24 percent of the labor force, and the group of workers over forty-four had shrunk to less than 31 percent (U.S.,

TABLE 9.1
Injury Frequency Rates, Compensation Cost per Injury, and Compensation Cost by Age and Sex, 1970

	Injury Frequency per Unit of Exposure	Compensation Cost per Injury	Compensation Cost per Unit of Exposure
Males	12.0	$1,558	$18,732
Less than 25 years old	22.3	789	17,585
25 to 44 years old	10.9	1,421	15,506
45 years old and over	10.9	2,072	22,550
Females	5.2	1,176	6,160
Less than 25 years old	4.5	557	2,501
25 to 44 years old	4.9	1,169	5,693
45 years old and over	5.8	1,379	8,057

Source: Author's calculations.

Notes: A unit of exposure is five hundred workers.

Injury frequency × compensation cost per injury = compensation cost per unit of exposure.

Department of Labor 1980a). Labor force projections for the near future predict that the groups of both youngest and oldest workers will shrink while the prime-age group will grow as a share of the labor force (Fullerton 1980). The age distribution changes between 1960 and 1979 should have had the effect of increasing the number of injuries experienced by the economy.

Demographic changes in the composition of the labor force have been dominated by the increased labor force participation of women. In 1960 women constituted little more than 33 percent of the labor force, but by 1979 their share was up to almost 42 percent (U.S., Department of Labor 1980a). According to perceived sex differences in injury frequency, such a change in labor force composition would reduce the annual total number of injuries for the economy. The limited evidence on sex-specific injury rates suggests that males have injury frequency rates exceeding those of females (see, e.g., Oi 1974 and Dillingham 1981b). One recent study has revealed that much, if not all, of this difference in injury experience is due to the occupational distribution of the sexes (Dillingham 1981b). But since the occupation classification used below is relatively aggregated, it is important to control for the sex composition of the labor force in order to help capture both the effect of a changing occupational mix over time and any changes in injury experience caused by true sex differences in injury frequency.

Injury frequency rates for three age groups by sex are displayed in the left-hand column of table 9.1. Two strong patterns are exhibited: the much higher frequency rates for males relative to females and an injury rate for young males that is twice that for both groups of older males. These rates reveal the counterbalancing effect of demograhic changes in the labor force between 1960 and 1980: injury frequency is pushed up by the increased share of young workers but pulled down by the increase in female relative to male workers.

The nature of work and the workplace also affects injury frequency and severity. Consequently, the industry and occupation mix of the economy will influence the incidence of injuries. The pattern of injury frequency by industry is well known; data on injuries by industry category have been collected by the federal government for years. The production of goods is more hazardous than production of services. Injury frequency rates are highest in construction, mining, manufacturing, transportation, communications, and utilities; they are lowest in the trade and service sectors (U.S., Department of Labor 1979b). Between 1960 and 1979, the industry composition of employment shifted toward the latter sectors and away from the former sectors, thus tending to depress economywide injury frequency.

Similar patterns exist across occupations: blue-collar workers face a great deal more risk of workplace injury than do white-collar workers. Evidence on occupation-specific injury risk comes primarily from my earlier work (Dillingham 1979). As could be expected, there is much greater variation in risk by occupation than by industry. Unskilled laborers with an injury frequency rate of 50.5 face twenty-five times the risk that professional and technical workers face on average, 2.0. Blue-collar workers, especially unskilled ones, face the greatest risk when employed in manufacturing. The injury rate for such workers is about 80.

Since 1960 the occupational distribution of employment has shifted so that white-collar workers, especially skilled ones, make up a larger portion of the labor force. The share accounted for by service workers has also increased, and both occupational groups have grown at the expense of blue-collar occupations. So the industry and occupation distributions have also altered in a way that reduces overall injury frequency.

Recall the changes in the age and sex distributions: the only change tending to increase frequency is the reduced age of the labor force. The other changes tend to reduce injury frequency, and the effect of all three may outweigh the age effect.

A Model of Compensation Cost

Information on employment and injury risk is insufficient for determining the relationship between labor force composition and workers' compensation costs. The differences in age-specific injury rates, for example, reveal nothing about the effect on costs of the changing age distribution of the work force. In order to derive aggregate compensation cost per unit of exposure, an estimate of average compensation cost per injury for each age, sex, industry and occupation group is necessary.

Since variation in the average compensation cost should reflect only group characteristics that influence compensation costs, cost data from only one workers' compensation system is utilized. This approach permits the specific benefit provisions of the law to be constant. Estimates of cost per injury will then reflect differences in wage levels, average injury severity, and the personal characteristics of each group. To eliminate the confounding effects of changing workers' compensation laws over time, these estimates are based on workers' compensation data for one year. The average compensation cost estimates are thus constants in the analysis. They are proxies for group differences in average compensation cost per injury. These estimates are used with the injury rate data to compute the unit cost estimates.

In practice, a linear model of workers' compensation cost such as that described is very difficult to construct. Benefit provisions are complex and vary by state, and they have changed over time. Measures of injury rates are generally unavailable at almost any disaggregated level. Total annual benefit payments reported by workers' compensation systems include payments for injuries incurred in previous years. The variation in and complexity of these plans precludes the direct use of inferential statistical techniques in pursuing the objectives of this study. A second-best approach is therefore adopted. This model employs a unique set of injury risk and compensation cost data. Workers' compensation and census data for one state at one point in time are used. Following are descriptions of data sources and construction of each of the three components.

Injury Rates. Injury experience or probability of injury is measured as an injury frequency rate, the number of injuries per million employee hours, or per five hundred full-time workers.[1] Data from New York State's workers' compensation system have been used in conjunction with 1970 census data to calculate age-, sex-, industry- and occupation-specific injury rates for the 1970 New York work force. Individual case records from accidents occurring in 1970 are used to calculate the number of injuries for each cell.[2] Then as a measure of exposure to injury risk, estimates of annual hours worked for each cell are derived for the New York work force from the 1970 census.[3] An injury frequency rate is simply the ratio of injuries (times one million) to exposure.

The employment cells for which both injury rates and compensation costs are calculated are relatively disaggregated: a matrix 2 by 3 by 8 by 27, or 1,296 cells. The matrix consists of two sex and three age

1. This equivalence assumes that year-round, full-time workers are employed an average of 2,000 hours a year.

2. These case records include instances of occupational diseases. However, the number of such cases is small — about 1 percent — and these cases generally involve the older, traditional occupational diseases, e.g., skin problems. Consequently, this study excludes from consideration the newer, more severe, long latency diseases.

3. The actual procedures employed to construct these very disaggregated injury rates were quite complicated. They are fully described in Dillingham (1979). Major issues are noted here.

The injury data from the New York Workmen's Compensation Board (WCB), the administrative agency for the state's system, are comprised of injury cases reported through Workers' Compensation *and* settled by the WCB. The WCB tallies the number of accidents reported to the system each year. During the 1960s and the first half of the 1970s, the annual number of accidents reported averaged 666,000. Of these reported accidents, only about 25 percent were actually reviewed by the WCB as potentially compensable. The accidents classified as new cases assembled averaged about 175,800 annually over the 1960–75 period. The injury data from which the injury rates are estimated are from the set of closed compensated cases, i.e., all of the new cases assembled that were determined to be compensable and were ultimately settled by the WCB.

In the late 1960s, these cases numbered about 115,000 annually. Included were not only physical injuries but instances of occupational diseases, as well as cases involving

groups, eight major occupation groups, and twenty-seven industry categories.[4] The injury frequency rates in table 9.1 are aggregated calculations from the same sources as those used to calculate the rates for the more detailed matrix.

Compensation Costs. Compensation costs are defined as the dollar awards paid in injury cases for medical expenses, lost wages, and disability or death awards. This model does not use data on premium costs incurred by employers because the characteristics of compensation systems are complex and vary across states, and it is therefore impossible to isolate demographic or labor market influences on workers' compensation costs. The two definitions of costs are directly related over the long run, and predictions about changes in costs using one measure imply changes in costs using the other. Cost data are taken from the same file of New York cases for 1970 as was used for the injury rate calculation. These case records contain all indemnity costs, but they do *not* contain information on medical costs. The indemnity award, however, reflects injury severity (e.g., permanent as opposed to temporary disability) and amount of lost work, both of which are positively correlated with medical cost. The award also depends upon the worker's wage. Average compensation per injury is calculated for each cell of the age, sex, industry and occupation matrix.[5]

---

third-party legal actions. It should also be noted that worker coverage by New York State's system is comprehensive and that there is no reason to believe that the system's specific characteristics bias the injury-rate estimates relative to estimates from another state system.

The census data are employed to obtain measures of exposure to risk. Since the 1970 occupation and industry affiliation of each census individual is known, annual hours worked can be calculated for each occupation-industry cell (controlling for age and sex) as the product of weeks worked per year and hours worked per week.

This exposure measure is really an estimate, since the census variable "hours worked" refers to the 1970 reference week for the census and the "weeks worked" variable is drawn from 1969 work experience. Changes in economic conditions between 1969 and 1970 suggest that use of the data for 1969 weeks worked might bias upward the exposure measure by no more than 1 percent, which would give a downward bias to the injury rate measures.

4. These age groups are less than twenty-five years old, twenty-five to forty-four years old, and over forty-four. Occupation and industry categories are available from the author.

5. The statistical significance of these estimates is explored in an appendix available from the author.

The product of injury rates and compensation cost per injury yields compensation cost per unit of exposure. Once unit cost is determined for each cell of the employment matrix, all that is required to calculate aggregate compensation cost for an economy is the distribution of employment for a particular year.

Because the workers' compensation system characteristics, wages and prices, and injury rates are being held constant at 1970 levels for New York State, the changes in compensation cost per unit of exposure from 1960 to 1980 — the object of this study — will be brought about *only* by changes in the age, sex, industry and occupation distributions of employment.

Employment Data. The simulation is performed by choosing realistic, meaningful changes in labor force composition, inserting them in the model as employment shares, and calculating the change in system costs. Thus one could use data for the United States, or individual states (to account for interstate variations in compensation costs), or projected future changes in the labor force (to predict future changes in compensation costs). The analysis here is restricted to actual and projected changes in the United States labor force, 1960 to 1978 and 1985 to 1995.

To obtain employment data for 1960, 1970, and 1978 (the latest year available), two data sources were used. For 1960 and 1970, the Census of Population was used (the One-in-One-Thousand Public Use Sample), and for 1978, the March 1978 Current Population Survey was used. For each year employment for each cell of the matrix was obtained, and cell employment as a share of total employment calculated.[6]

Empirical Results

Compensation Costs by Age and Sex. In table 9.1 injury rates and compensation cost per injury and per unit of exposure are provided for each age-sex cell in the analysis. For each age group, male compensation cost per injury exceeds that for females. But the greatest cost variation occurs across age groups, and the pattern for

---

6. Confidence intervals for these calculations are reported in an appendix available from the author.

TABLE 9.2
Compensation Cost
per Unit of Exposure
by Age, Sex, and Occupation, 1970

	Professional, Technical, Managers, Administrators, Sales Workers	Clerical Workers	Craftsmen
Males			
Less than 25 years old	$2,403	$8,023	$19,512
25 to 44 years old	3,240	6,109	21,168
45 years old and over	4,758	7,487	32,786
Females			
Less than 25 years old	2,093	935	8,385
25 to 44 years old	2,932	2,213	4,367
45 years old and over	5,413	4,518	11,817

	Operatives	Laborers	Service Workers
Males			
Less than 25 years old	$30,680	$52,829	$13,549
25 to 44 years old	30,419	68,884	14,184
45 years old and over	42,695	98,233	21,401
Females			
Less than 25 years old	9,316	4,353	6,921
25 to 44 years old	14,150	9,310	15,439
45 years old and over	11,675	6,442	18,711

Source: Author's calculations.

males and females is similar: costs for workers less than twenty-five years old average about 40 percent of the costs of workers over forty-four. Again, it is important to recognize that these cost patterns cannot be attributed entirely to age or sex. The age and sex differences in both injury frequency and injury cost reflect in part differences in the occupation and industry distributions. An extreme case of this confusion of variables exists in the female age–injury rate relationship. If age-specific differences in occupation distribution are controlled for, the female age–injury rate relationship becomes negative, just like the male pattern (Dillingham 1981a). In general, however, adjustment for these other influences does not alter the key result for compensation cost per unit of exposure.

Table 9.1 contains the product of the injury rate and compensation cost per injury: compensation cost per unit of exposure. Two patterns are revealed. One is the male-female cost differential. On average, male costs per unit of exposure are three times those of females. The second pattern is the apparent positive relationship between age and cost per unit of exposure. Both patterns are highlighted in table 9.2 where cost per unit is broken down by age, sex, and major occupation category. For males, there is a strong age profile to compensation costs. The age profile is less clearcut for women, except in the white-collar and service occupations.

In the blue-collar sector, the sex differential in cost is most dramatic. There are at least two reasons for such a differential. Within an occupation group there is great variety in actual work performed, and so women in that group are employed in fundamentally different jobs and experience different kinds of risk and injuries. Second, since these costs are in part determined by wage levels, a female-male wage ratio of less than one implies lower compensation costs for women.

In summary, compensation cost per unit of exposure is higher for males than females, and compensation cost per unit of exposure tends to increase as cohort age increases, although this relationship varies somewhat by sex and occupation category.

Compensation Costs by Industry and Occupation. The frequency, severity, and nature of workplace accidents depend upon both production techniques and the bundle of goods and services produced. Therefore, the compensation cost per unit of exposure should vary across occupations and industries. Estimates of cost per unit of exposure for major industries and occupations are provided in table 9.3.[7] For each industry-occupation cell, estimates for all workers and for prime-age males are provided. These estimates demonstrate an unambiguous phenomenon: the extraction of natural resources, construction, and manufacture and transportation of economic goods generate the highest cost per unit of exposure. In addition, the more directly in-

7. Some industry and occupation groups in the full model have been excluded from these tables. These groups include agriculture and services, and government categories as well as farmer and domestic service worker occupations. They are excluded because the rate and cost calculations are not necessarily representative of all members of these industry and occupation groups. For examples, many agricultural workers were not covered by workers' compensation in 1970, and some government employees are included in other occupation and industry groups.

TABLE 9.3
Compensation Cost
per Unit of Exposure
by Major Industry and Occupation Group,
1970

	Professional, Technical, Managers, Administrators, Sales Workers	Clerical Workers	Craftsmen
Construction and mining			
All workers	$6,534	$2,638	$48,248
Males, 25–44	2,768	3,670	36,943
Manufacturing, nondurable			
All workers	4,299	5,990	17,810
Males, 25–44	4,298	12,750	14,131
Manufacturing, durable			
All workers	2,226	3,683	19,418
Males, 25–44	2,235	5,912	16,775
Transportation			
All workers	5,454	4,325	26,954
Males, 25–44	6,544	5,509	24,625
Communication and utilities			
All workers	3,129	2,711	12,524
Males, 25–44	5,489	9,186	16,402
Wholesale and retail trade			
All workers	5,948	5,997	24,866
Males, 25–44	4,815	9,716	21,871
Service industries			
All workers	2,815	2,987	18,562
Males, 25–44	2,370	2,875	12,984
Aggregate occupation			
All workers	3,846	4,027	25,548
Males, 25–44	3,240	6,109	21,168

volved in these activities is the job assignment, the higher the per unit of exposure cost. Blue-collar employment generates the highest compensation cost per unit of exposure, and within this occupational sector the least skilled work generates the highest costs. In New York, blue-collar employment accounts for 31 percent of total employment in 1970, 71 percent of all injuries, and 72 percent of total compensation costs (Dillingham 1979).

Economywide Compensation Costs, 1960 to 1980. As the composition of employment changes, differences in compensation costs per unit of exposure for the age, sex, and industry and occupation groups will

TABLE 9.3 (continued)

	Operatives	Laborers	Service Workers	Aggregate Industry
Construction and mining				
All workers	$66,536	$128,988	$50,844	$48,654
Males, 25–44	47,592	115,385	9,732	40,371
Manufacturing, nondurable				
All workers	28,488	126,340	19,839	18,867
Males, 25–44	43,101	191,619	16,259	22,452
Manufacturing, durable				
All workers	29,001	92,982	15,201	17,969
Males, 25–44	29,323	91,388	22,462	16,605
Transportation				
All workers	35,282	75,519	22,572	27,016
Males, 25–44	28,295	60,737	50,249	26,178
Communication and utilities				
All workers	14,203	18,452	2,282	7,611
Males, 25–44	10,733	29,479	2,245	11,071
Wholesale and retail trade				
All workers	29,870	40,982	14,439	12,554
Males, 25–44	28,157	58,952	11,599	13,150
Service industries				
All workers	24,165	67,679	20,060	8,026
Males, 25–44	32,718	57,477	18,655	7,562
Aggregate occupation				
All workers	29,508	72,361	16,757	. . .
Males, 25–44	30,419	68,884	14,184	. . .

Source: Author's calculations.

cause changes in aggregate economywide compensation cost per unit of exposure.

Using this model, two variables are needed to calculate economywide compensation cost per unit of exposure: the compensation cost per unit of exposure for age, sex, industry and occupation groups and the relative share of total employment accounted for by each group. The patterns in the compensation cost variable and the broad trends in employment, 1960–80, have been described. Based on the compensation cost estimates and employment trends since 1960, economywide compensation cost per unit of exposure should decrease between 1960 and 1980.

Data from the detailed employment matrix (1,296 cells) are used to estimate the extent of the change in economywide cost per unit of exposure between 1960 and 1978. The results are presented in

TABLE 9.4
Changes in Economywide
Compensation Cost per Unit of Exposure,
1960–95

	Model A[1]	Model B[2]	Aggregated Model A[3]
1960	$19,974	$20,732	$14,501 (actual)
1970	17,056	17,160	13,669 (actual)
1978	16,254	16,306	13,005 (actual)
1985			12,436 (projected)
1990			12,261 (projected)
1995			12,302 (projected)

Sources: Actual estimates are based on data from New York's workers' compensation system, the 1960 and 1970 Current Population Survey, and the March 1978 Current Population Survey. Projected estimates are based on injury rates and compensation cost data, table 9.3, and the projected age-sex share of the labor force, 1985–95. See Fullerton (1980).

Notes: 1. Employment shares reflect only employment data.

2. Employment shares calculated from employment data adjusted for age/sex differences in hours worked.

3. Differences in 1960–1978 estimates between the first and third column arise because the projections required a more aggregated model. See text for further discussion.

the first two columns of table 9.4. Because relative employment shares were calculated in two ways, two sets of estimates are presented. Model A uses only employment data for calculating employment shares. In model B employment shares were adjusted for age and sex differences in median weekly hours worked. Since young workers and women tend to work fewer hours than others, omission of the hours adjustment increases their relative shares of employment.

As expected, the economywide estimates do decline by roughly 20 percent over the period. Most of the decrease in cost, about 80 percent, occurs between 1960 and 1970. By holding constant the age and sex distribution of employment and varying the industry and occupation distribution, and vice versa, the major causes of the decrease in costs can be identified. Increased numbers of young and female workers and decreased numbers of older workers exert steady downward pressure on costs over this twenty-year period. Changes in the industry and occupation mix of the economy have their biggest influence on costs between 1960 and 1970. The industry-occupation effect is stronger than the demographic effect in this earlier period.

The reason that most of the cost decrease occurs in the 1960–70 period is that there is both a demographic and economic (industry-occupation) effect in the earlier decade, but in the later decade the economic effect is greatly reduced.

Projected Changes in Economywide Compensation Cost. Demographic and economic change in the 1960 to 1980 period caused compensation cost per unit of exposure to decrease substantially. Will this trend continue into the future? If one is willing to make projections about future changes in age and sex composition and industry-occupation mix of the labor force, this question can be addressed with the same model employed to estimate changes in past costs.

In this exercise a more aggregated version of the earlier model is used. Data on injuries, exposure, and injury costs are aggregated across industries and occupations to obtain injury rates and average compensation cost per injury for the six age and sex groups identified in table 9.1. This aggregation is used because labor force projections are available only by age and sex. Given the nature of the data used in calculating injury rates and compensation costs and given the level of aggregation of the labor force projections, this simulation implicitly maintains the industry-occupation employment distribution for the six age and sex groups at 1970 levels.

The economywide estimates for 1960–78 actual compensation costs based on the more aggregated version of the model are presented in table 9.4 along with estimates of projected costs for 1985, 1990, and 1995. The projected estimates are based on the middle growth assumption of the Bureau of Labor Statistics (see Fullerton 1980). The 1960, 1970, and 1978 estimates are lower than those for models A and B because the aggregation across industries and occupations eliminates extremely high-cost cells, thus reducing the weighted mean costs.

Projected economywide compensation cost decreases from 1978 to 1990, but begins to increase by 1995. This turnabout in compensation cost occurs for a combination of reasons. The force that tends to depress costs throughout this period is the increased participation of women in the labor force (or the increased share of employment accounted for by jobs that have historically been filled by women). Recall the large sex differential in compensation cost per

TABLE 9.5
Workers' Compensation
Medical Costs by
Age Group

Age Group	Average Incurred Medical Cost[1]	Medical Cost, Adjusted to 1971 Prices[2]	Indemnity Costs[3]
Less than 25 years	$ 760	$392	$ 753
25 to 44 years	1,086	560	1,385
44 years and over	1,331	686	1,711

Notes: 1. Calculated from data for 1979–80 claims provided by the National Council on Compensation Insurance.

2. Medical costs are deflated to 1971 price levels, using the medical care component of the Consumer Price Index.

3. Compensation costs are derived from workers' compensation sources described in the text.

unit of exposure. This depressing force is weaker by 1995, however, because labor force projections incorporate a slowed rate of increase in female labor force participation by 1995.

The highest cost group of the six age and sex groups identified here is the male group over 45 years of age. This group's share of the labor force declines up to 1990, depressing economywide costs; but after 1990, labor force age distribution shifts toward the older groups — both males and females — and thus compensation costs are pushed higher. In summary, the turnaround in compensation cost trends by 1995 occurs because the high-cost demographic groups' portion of the labor force grows and that of the low-cost groups shrinks.

Interpretation of Findings

A hypothetical situation has been analyzed: if workers' compensation system characteristics, wages, prices, and group injury frequency and severity are held constant, what would happen to economywide compensation cost per unit of exposure over the 1960–80 period as demographic change occurred in the labor market and the industry-occupation mix of the economy changed? Given these assumptions and the variation in compensation cost by age, sex, industry, and occupation that has been described, compensation cost per unit of exposure decreases between 1960 and 1980.

This result should be qualified in several ways. First, because of data limitations, medical costs have been ignored. They should, however, be positively correlated with injury severity, and their inclusion would therefore reinforce the results of the study. Limited evidence from the National Council on Compensation Insurance (NCCI) supports this proposition. Average medical cost incurred by claimant age group can be computed from NCCI's detailed claim call sample of claims for the 1979–80 period.[8] This sample is based on data from twelve states, one of which is New York. The New York cost data for the three age groups used throughout this study are reproduced in table 9.5. They have been deflated to the 1971 price level in order to compare them with indemnity costs, which reflect 1970–71 price and wage levels. Although medical costs do rise with age, the age profile of medical costs is less steeply sloped than that for indemnity costs.

Second, the workers' compensation injury data suffer from a small, but important bias. Injury cases which were compensated but which required several years to close are underreported. These are serious injury cases, which are concentrated among older workers (see Dillingham 1979). Thus, the reported age differential in compensation cost is reduced, and the effect of a changing age distribution on economywide compensation cost per unit of exposure is therefore biased downward.

Third, changes in wages induced by the market forces that brought about the economy's structural change over this period have also been ignored. Wage changes have a direct effect on compensation costs. If sectoral shifts in the economy occur in response to changes in demand, then earnings in growth sectors should increase relative to earnings in declining sectors. Such a pattern in relative wage changes would tend to push up compensation costs and offset the extent of the decrease estimated here. Limited evidence, however, suggests that such wage effects on compensation costs are negligible. The age-sex wage structure seems to have changed over the 1960s and 1970s in such a way as to reinforce the findings of this study. The increase in

8. These data were obtained from NCCI through the Alliance of American Insurers. It should be noted that these data are based on relatively early insurance carrier claim evaluation and are subject to revision. Inferences from these data should be considered preliminary.

the number of young people, especially males, in the labor market in the late 1960s and early 1970s depressed the earnings of young males relative to older males and had no effect on the age cohort earnings of women (Freeman 1979). Such a shift in male earnings by age group would amplify the negative effect on compensation cost of the demographic changes since 1960. Furthermore, a study by Freeman (1980) of changes in the occupation and industry distribution of employment between 1960 and 1970 found that changes in occupation-industry employment patterns occurred with relatively little corresponding change in relative incomes of occupation-industry categories. It seems, therefore, that any bias in these results that could be attributed to changes in relative wages would be small.

A final warning is appropriate. The estimates of compensation cost per unit of exposure show that older workers have higher costs than younger workers and that male workers have higher costs than females. These results in no way imply that a younger or female worker will cost an employer less than an older or a male worker. These employment groups are very aggregated, and therefore, include a great deal of heterogeneity in work and workplace. At this level of aggregation it is difficult, if not impossible, to determine how much of the age-sex differential in cost is a worker effect and how much an industry or occupation effect.

Conclusions and Implications

There are five major findings in this study:

1. Compensation cost per unit of exposure (million employee hours) varies markedly by age and sex group, as well as by industry and especially occupation category.

2. The high-cost groups for each of the four categories are older workers; males; construction, mining, and manufacturing; and blue-collar workers, especially unskilled labor.

3. An undetermined amount of the age and sex differences in cost reflects variation in industry and occupation that is unaccounted for because of the aggregated categories used in the study.

4. Variation in compensation cost per unit of exposure across these broad demographic groups and economic sectors implies that

changes in the structure of the economy can induce changes in the social cost of workers' compensation.

5. Actual changes in the structure of the economy between 1960 and 1980 have reduced economywide workers' compensation cost per unit of exposure. Specifically, these are the decreased share of the labor force accounted for by males and older workers in general and the decreased share of employment accounted for by blue-collar jobs in manufacturing and construction. Their downward pressure on unit cost is offset by inflation, money wage increases, and benefit liberalization — all of which increase average compensation cost per injury and therefore cost per unit of exposure. In the 1970–78 period, structural changes in the economy depressed estimated economywide unit compensation cost by about 5 percent. But simultaneously, wages and prices were rising on an average of more than 5 percent, and benefit reform increased benefit payments for any given level of wages. Also, of course, total social costs for workers' compensation tend to increase over time because the increasing size of the labor force offsets possible decreases in cost per unit of exposure. Thus, the net change in unit cost in the 1970s was very probably positive.

These conclusions have an important implication for future trends in workers' compensation costs. The increase in costs during the 1970s, attributed to inflation and benefit reform, occurred as demographic and economic change reduced cost per unit of exposure. If the latter changes had not occurred, the cost increases in the 1970s would have been even greater. Furthermore, projected demographic changes in the labor force of the 1990s suggest higher workers' compensation costs even if inflation and benefit reform come to a grinding halt in the 1980s. The 1990s cost increases predicted by this model could, of course, be offset by technological change or changes in industry-occupation distribution that reduce injury frequency or severity. If in the future, however, benefits continue to be liberalized or the compensation system is burdened with costly occupational disease claims, society may not be aided in financing the additional costs, as it was from 1960 to 1980, by offsetting demographic and economic trends.

These conclusions also reveal more about the environment in which social concern for, and legislation aimed at, increased safety developed in the 1960s. Economists have argued that the increase in injury frequency in the 1960s was due in large part to increased economic activity and the change in the age composition of the labor force during this period, not some exogenous increase in workplace injury risk (see Chelius 1979 and Smith 1974). Concern that workplace safety was inadequate and possibly worsening helped proponents of safety legislation enact the 1970 Occupational Safety and Health Act. The results presented here suggest that this cyclical increase in injury frequency and the heightened concern for occupational safety occurred at the same time as the development of strong secular forces that would reduce the unit cost of compensation and possibly the unit social cost of workplace accidents. The possibility that injury costs can vary inversely with injury frequency is clearly demonstrated by the analysis under Empirical Results.

The findings of this study are tentative only. The simulation exercise uses very aggregated data derived primarily from one year's workers' compensation experience in one state. The results reflect, therefore, the unique conditions of that time and place; the validity of these results would be vastly improved by the use of a multistate, multiyear model. But these tentative findings do suggest other research questions. What factors account for the large age and sex differentials in injury rates and compensation cost per injury? What factors explain the observed variation in compensation costs, differences in injury severity or differences in earnings? — and to what extent do the specific provisions of the law influence the answer to this question?

REFERENCES

Akerlof, George A., and Dickens, William T.

1982. "The Economic Consequences of Cognitive Dissonance." *American Economic Review* 72, 3: 307–19.

Allen, Steven G.

1981. "Compensation, Safety and Absenteeism: Evidence from the Paper Industry." *Industrial and Labor Relations Review* 34, 2 (January 1981): 207–18.

American Medical Association

1971. *Guides to the Evaluation of Permanent Impairment.* Chicago: AMA.

Antos, Joseph R.

1983. "Union Effects on White-Collar Compensation." *Industrial and Labor Relations Review* 36, 3 (April 1983): 461–479.

Becker, Gary

1971. *Economic Theory.* New York: Alfred A. Knopf.

Berkowitz, Monroe

1971. "Allocation Effects of Workmen's Compensation." *IRRA Proceedings.* Madison, Wis.: Industrial Relations Research Association, pp. 342–49.

1973a. "Income Replacement Benefits: 1." In *Compendium on Workmen's Compensation.* Washington, D.C.: Government Printing Office.

1973b. "Workmen's Compensation Income Benefits: Their Adequacy and Equity." In *Supplemental Studies for the National Commission on State Workmen's Compensation Laws,* vol. 1. Washington, D.C.: Government Printing Office.

Berkowitz, Monroe; Burton, John F., Jr.; and Vroman, Wayne

1979. *Final Report,* from the research project "An Evalution of State Level Human Resource Delivery Programs: Disability Compensation." Washington, D.C.: National Science Foundation.

Berkowitz, Monroe, and Johnson, William G.
1974. "Health and Labor Force Participation." *Journal of Human Resources* 9, 1 (Winter 1974): 117–28.

Brown, Charles
1980. "Equalizing Differences in the Labor Market." *Quarterly Journal of Economics* 94, 1 (February 1980): 113–34.

Burton, John F., Jr.
1978. "Wage Losses from Work Injuries and Workers' Compensation Benefits: Shall the Twain Never Meet?" In *1978 Convention Proceedings of IAIABC.* Quebec City, Quebec: International Association of Industrial Accident Boards and Commissions, pp. 74–83.

Burton, John F., Jr., and Berkowitz, Monroe
1971. "Objectives Other than Income Maintenance for Workmen's Compensation." *Journal of Risk and Insurance* 38, 3 (September 1981): 343–55.

Burton, John F., Jr.; Larson, Lloyd W.; and Moran, Janet P.
1980. "The Final Report on A Research Project on Permanent Partial Disability Benefits." Prepared for a joint labor-management committee of the State of New York. Ithaca, N.Y.: New York State School of Industrial and Labor Relations.

Burton, John F., Jr., and Vroman, Wayne
1979. "A Report on Permanent Partial Disabilities under Workers' Compensation." In *Research Report of the Interdepartmental Workers' Compensation Task Force* 6 (1979): 11–77.

Butler, Richard J.
1979. "Black/White Wage and Employment Changes: A Look at Production Workers in South Carolina, 1940–1971." Ph.D. dissertation, University of Chicago.
1981. "A Note on Hedonic Techniques in Economics." Unpublished paper.

Butler, Richard J., and Worrall, John D.
1981. "Workers' Compensation: Benefit and Injury Claims Rates in the Seventies." Mimeo. New York: National Council on Compensation Insurance.
1982. "Hazard Rate for Workers' Compensation." Paper presented at the second annual Seminar on Economic Issues in Workers' Compensation. National Council on Compensation Insurance, City University Graduate School and University Center, New York, N.Y.
1983. "Workers' Compensation: Benefit and Injury Claims Rates in the Seventies." *Review of Economics and Statistics.* Forthcoming.

Chamber of Commerce of the United States

1972–75. *Analysis of Workmen's Compensation Laws.* Annual editions. Washington, D.C.: Chamber of Commerce of the United States.

1978. *Employee Benefits 1977.* Washington, D.C.: Chamber of Commerce of the United States.

Cheit, Earl F.

1962. *Injury and Recovery in the Course of Employment.* New York: John Wiley and Sons.

Chelius, James R.

1973. "An Empirical Analysis of Safety Regulation." In *Supplemental Studies for the National Commission on State Workmen's Compensation Laws,* vol. 3, edited by Monroe Berkowitz. Washington, D.C.: Government Printing Office.

1974. "The Control of Industrial Accidents: Economic Theory and Empirical Evidence." *Law and Contemporary Problems* 38, 4 (Summer-Autumn 1974): 700–729.

1977. *Workplace Safety and Health: The Role of Workers' Compensation.* Washington, D.C.: American Enterprise Institute.

1979. "Economic and Demographic Aspects of the Occupational Injury Problem." *Quarterly Review of Economics and Business* 19 (Summer 1979): 65–70.

1982. "The Influence of Workers' Compensation on Safety Incentive." *Industrial and Labor Relations Review* 35, 2 (January 1982): 235–42.

Coase, Ronald

1960. "The Problem of Social Cost." *Journal of Law and Economics* 3 (October 1960): 1–44.

Devers, William J., Jr.

1981. "Praise for Wage-Loss System." In *IAIABC Journal* (January 1981): 9–10.

Dillingham, Alan

1979. "The Injury Risk Structure of Occupations and Wages." Ph.D. dissertation, Cornell University.

1981a. "New Evidence on Age and Workplace Injuries." *Industrial Gerontology* 4 (Winter 1981): 1–10.

1981b. "Sex Differences in Labor Market Injury Risk." *Industrial Relations* 20, 1 (Winter 1981): 117–22.

Dorsey, Stuart

1982. "A Model and Empirical Estimates of Worker Pension Coverage in the U.S." *Southern Economic Journal* 49, 2 (October 1982).

Dorsey, Stuart, and Walzer, Norman

1979. "Occupational Risk, Compensating Differentials, and Liability Rules." Mimeo. Western Illinois University.

1983. "Compensating Differentials and Liability Rules." *Industrial and Labor Relations Review.* Forthcoming 1983.

Ehrenberg, Ronald G., and Smith, Robert S.

1982. *Modern Labor Economics: Theory and Public Policy.* Glenview, Ill.: Scott, Foresman and Company.

Federal Employee Compensation Act

1976. 5 U.S.C. § 8101 (1976).

Fenn, Paul

1981. "Sickness Duration, Residual Disability, and Income Replacement: An Empirical Analysis." *Economic Journal* 91, 361 (March 1981): 158–73.

Filteau, Robert J.

1980. "When Stress Becomes Distress: Mental Disabilities under Worker's Compensation in Massachusetts." *New England Law Review* 15 (1979/1980): 287–307.

Flanagan, Robert J., and Mitchell, Daniel J. B.

1982. "Wage Determination and Public Policy." In *Industrial Relations Research in the 1970s: Review and Appraisal,* edited by Thomas A. Kochan, Daniel J. B. Mitchell, and Lee Dyer. Madison, Wis.: Industrial Relations Research Association, pp. 45-94.

Freeman, Richard B.

1979. "The Effect of Demographic Factors on Age-Earnings Profiles." *Journal of Human Resources* 14, 3 (Summer 1979): 289–318.

1980. "An Empirical Analysis of the Fixed Coefficient 'Manpower Requirements' Model, 1960–1970." *Journal of Human Resources* 15, 2 (Spring 1980): 176–99.

1981. "The Effect of Unionism on Fringe Benefits." *Industrial and Labor Relations Review* 34, 4 (July 1981): 489–509.

Fullerton, Howard N., Jr.

1980. "The 1995 Labor Force: A First Look." *Monthly Labor Review* 103, 12 (December 1980): 11–21.

Ghez, Gilbert R., and Becker, Gary S.

1975. *The Allocation of Time and Goods over the Life Cycle.* New York: National Bureau of Economic Research, pp. 102–20.

Gross, Keith H.

1977. "New York's Workmen's Compensation Law and the Concept of Accidental Injury." *Syracuse Law Review* 28: 664–99.

Heckman, James J.

1979. "Sample Selection Bias as a Specification Error." *Econometrica* 47, 1 (January 1979): 153–60.

Interdepartmental Workers' Compensation Task Force

1977. *Workers' Compensation: Is There a Better Way? A Report to the President and the Congress on the Need for Reform of State Workers' Compensation.* Olympia Fields, Ill.: Workers' Disability Income Systems, Inc.

Johnson, William G.; Cullinan, P. R.; and Curington, W. P.

1979. "The Adequacy of Workers' Compensation Benefits." *Research Report of the Interdepartmental Workers' Compensation Task Force,* vol. 6, pp. 95–122.

Kalachek, E., and Raines, F.

1980. "Trade Unions and Hiring Standards." *Journal of Labor Research* 1, 1 (Spring 1980).

Keefe, Steve

1981. "Wage Loss: Panacea or Pandora's Box?" Part I. In *IAIABC Journal* (June 1981): 8–9.

Kossoris, Max D.

1940. "Relation of Age to Industrial Injuries." *Monthly Labor Review* 51, 4 (October 1940): 789–804.

Lambrinos, James

1981. "Health: A Source of Bias in Labor Supply Models." *Review of Economics and Statistics* 63, 2 (May 1981): 206–12.

Larson, Arthur

1970. "Mental and Nervous Injury in Workmen's Compensation." *Vanderbilt Law Review* 23, 6 (November 1970): 1243–75.

1973. "Basic Concepts and Objectives of Workmen's Compensation." In *Supplemental Studies for the National Commission on State Workmen's Compensation Laws,* vol. 1. Washington, D.C.: Government Printing Office, pp. 31-39.

1978. *The Laws of Workmen's Compensation.* Four vols. and cumulative supplement. New York: Matthew Bender.

1982. "The Current Status of the Compensation Exclusivity Doctrine." Paper presented at the NCCI Seminar on Workers' Compensation Current Trends, New York, N.Y.

Lasky, Herbert

1980. "Psychiatry and California Worker's Compensation Laws." *California Western Law Review* 17, 1 (Fall 1980): 1–25.

Makarushka, Julia Loughlin; Chollet, Deborah; and Frankel, Martin

1977. "The Health Studies Survey of Workers' Compensation Recipients in New York, Florida, Wisconsin, Washington, and California: Survey Design and Administration." Health Studies Program Working Paper no. 25. Syracuse, N.Y.: Syracuse University, Maxwell School.

Makarushka, Julia Loughlin, and Johnson, W. G.

1976. "The Experience of Injured Workers' Compensation: A Report to the Interdepartmental Task Force on Workers' Compensation." Health Studies Program Working Paper no. 15. Syracuse, N.Y.: Syracuse University, Maxwell School.

Mellow, Wesley S.

1982. "Employer Size and Wages." *Review of Economics and Statistics* 64, 3 (August 1982): 495–501.

Mitchell, Olivia S.

1982. "The Labor Market Impact of Federal Regulation: OSHA, ERISA, EEO, and Minimum Wage." In *Industrial Relations Research in the 1970s: Review and Appraisal,* edited by Thomas A. Kochan, Daniel J. B. Mitchell, and Lee Dyer. Madison, Wis.: Industrial Relations Research Association, pp. 149–86.

Nagi, Saad Z.

1969. *Disability and Rehabilitation.* Athens, Ohio: Ohio University Press.

1975. *An Epidemiology of Adulthood Disability in the United States.* Mershon Center Informal Publications. Columbus, Ohio: Mershon Center, Ohio State University.

Naples, Michele I., and Gordon, David M.

1981. "The Industrial Accident Rate: Creating a Consistent Time Series." Mimeo.

National Commission on State Workmen's Compensation Laws

1972. *The Report of the National Commission on State Workmen's Compensation Laws.* Washington, D.C.: Government Printing Office.

1973. *Compendium of Workmen's Compensation.* Washington, D.C.: Government Printing Office.

National Council on Compensation Insurance

1981. *ABC's of Experience Rating.* New York: NCCI.

1981. *The Pricing of Workers' Compensation Insurance.* New York: NCCI.

National Data Use and Access Laboratories

1973. *Technical Documentation for the 1960 Public Use Sample.* Washington, D.C.: Government Printing Office, January.

New York Workers' Compensation Board

1979. *Compensation Cases Closed, 1979.* New York.

Oi, Walter Y.

1973. "An Essay on Workmen's Compensation and Industrial Safety." In *Supplemental Studies for the National Commission on State Workmen's Compensation Laws,* vol. 1. Washington, D.C.: Government Printing Office, pp. 44–106.

1974. "On the Economics of Industrial Safety." *Law and Contemporary Problems* 38, 4 (Summer/Autumn 1974): 669–99.

Olson, Craig A.

1981. "An Analysis of Wage Differentials Received by Workers on Dangerous Jobs." *Journal of Human Resources* 16, 2 (Spring 1981): 167–85.

Price, Daniel N.

1978. "Workers' Compensation: Coverage, Benefits, and Costs, 1976." *Social Security Bulletin* 41, 3 (March 1978): 30–34.

1979. "Workers' Compensation Program in the 1970s." *Social Security Bulletin* 42, 5 (May 1979): 3–24.

1980. "Workers' Compensation: 1978 Program Update." *Social Security Bulletin* 43, 10 (October 1980): 3–10.

1981. "Workers' Compensation Coverage, Benefits, and Costs, 1979." *Social Security Bulletin* 44, 9 (September 1981): 9–13.

Reid, James J.

1966. "Report of the Permanent Partial Disabilities Committee." In *1966 Convention Proceedings of the IAIABC,* International Association of Industrial Accident Boards and Comissions: 27–71.

Report of the Full Compliance Subcommittee to Consider State Compliance with Workers' Compensation Recommended Standards

1976. 1st ed. and supplement. Washington, D.C.: Workers' Disability Income Systems, Inc.

Rosen, Sherwin

1974. "Hedonic Prices and Implicit Markets: Product Differentiation in Pure Competition." *Journal of Political Economy* 82 (1974): 34–55.

Russell, Louise B.

1973. "Pricing Industrial Accidents." In *Supplemental Studies for the National Commission on State Workmen's Compensation Laws,* vol. 3, edited by Monroe Berkowitz. Washington, D.C.: Government Printing Office.

1974. "Safety Incentives in Workmen's Compensation Insurance." *Journal of Human Resources* 9, 3 (Summer 1974).

Seifert, Mary Lee

1981. "Changes in the Spendable Earnings Series for 1981." *Employment and Earnings* 28, 3 (March 1981): 9–13.

Sherman, P. Tecumseh

1917. "The Basis of Compensation." In *Proceedings of the Conference on Social Insurance.* Bulletin 212. Washington, D.C.: U.S., Department of Labor, Bureau of Labor Statistics, pp. 184–92.

Skolnik, Alfred

1972. "Workmen's Compensation Payments and Costs, 1970." *Social Security Bulletin* 35 (January 1972).

Smith, Adam

1776. *The Wealth of Nations,* edited by R. H. Campbell. 2 vols. London: Oxford University Press, 1976.

Smith, Robert S.

1973. "An Analysis of Work Injuries in Manufacturing Industry." In *Supplemental Studies for the National Commission on State Workmen's Compensation Laws,* vol. 3, edited by Monroe Berkowitz. Washington, D.C.: Government Printing Office, pp. 10–26.

1974. "The Feasibility of an 'Injury Tax' Approach to Occupational Safety." *Law and Contemporary Problems* 38, 4 (Summer-Autumn 1974): 730–44.

1979a. "Compensating Wage Differentials and Public Policy: A Review." *Industrial and Labor Relations Review* 32, 3 (April 1979): 339–52.

1979b. "The Impact of OSHA Inspections on Manufacturing Injury Rates." *Journal of Human Resources* 14, 2 (Spring 1979): 145-70.

1981. "Protecting Workers' Health and Safety." In *Instead of Regulation,* edited by Robert W. Poole. Lexington, Mass.: Lexington Books, pp. 311–38.

Staten, Michael, and Umbeck, John

1982. "Information Costs and Incentives to Shirk: Disability Compensation of Air Traffic Controllers." *American Economic Review* 72, 5 (December 1982).

Thaler, Richard, and Rosen, Sherwin

1975 "The Value of Saving a Life: Evidence from the Labor Market." In *Household Production and Consumption,* edited by N. Terleckyj. New York: National Bureau of Economic Research.

Tinsley, La Verne C.

1982. "Workers' Compensation: Key Legislation in 1981." *Monthly Labor Review* 105, 2 (February 1982): 24–30.

U.S., Congress, House, Committee on Appropriations

1970. *Air Traffic Controllers* (Corson Report). H. Rept. 91-1012. 91st Congress, 2d sess., July 9.

1976. *Administration of the Federal Employee's Compensation Act.* H. Rept. 94-1757. 94th Congress, 2d sess., October 6.

1978. *Hearing on Department of Labor Appropriations for 1979.* Part 1, 95th Congress, 2d sess., February 6.

U.S., Department of Commerce, Bureau of the Census

1972. *Public Use Samples of Basic Records from the 1970 Census: Description and Technical Documentation.* Washington, D.C.: Government Printing Office.

1977. *Employers' Expenditures for Employee Compensation.* Washington, D.C.: Government Printing Office.

1978a. *Current Population Survey.* Washington, D.C.: Government Printing Office.

1978b. *Technical Documentation: Annual Demographic File (March Supplement of Current Population Survey) 1978.* Washington, D.C.: Government Printing Office.

1979. *Statistical Abstract of the United States: 1979.* 100th ed. Washington, D.C.: Government Printing Office.

U.S., Department of Labor

1982. *Monthly Labor Review* 105, 4 (April 1982).

U.S., Department of Labor, Bureau of Labor Statistics

1979a. *Employment and Earnings, States and Areas, 1939–1978.* Washington, D.C.: Government Printing Office.

1979b. *Occupational Injuries and Illnesses in the United States by Industry, 1976.* Bulletin 2019. Washington, D.C.: Government Printing Office.

1980a. *Handbook of Labor Statistics.* Bulletin 2070. Washington, D.C.: Government Printing Office.

1980b. *Occupational Injuries and Illnesses in 1978: Summary.* Report 586. Washington, D.C.: Government Printing Office.

1980c. *Occupational Injuries and Illnesses in the United States by Industry, 1974–1982.* Washington, D.C.: Government Printing Office.

U.S., Department of Labor, ESA-OWCP

1981. *State Compliance with the 19 Essential Recommendations of the National Commission on State Workmen's Compensation Laws, 1972–1980.* Washington, D.C.: Government Printing Office.

U.S., Department of Transportation, FAA

1979. *Report to Congress on the FAA Reevaluation of Second Career Training.* Washington, D.C.: Government Printing Office.

U.S., Department of Transportation, FAA, Office of Aviation Medicine

1978. *FAA-Aviation Medicine Report,* no. 78-21.

Victor, Richard B.

1982. "Experience Rating and Workplace Safety." Paper presented at the National Council on Compensation Insurance Seminar: Economic Issues in Workers' Compensation. New York, N.Y.

Viscusi, W. Kip

1979a. *Employment Hazards: An Investigation of Market Performance.* Cambridge, Mass.: Harvard University Press.

1979b. "Job Hazards and Worker Quit Rates: An Analysis of Adaptive Worker Behavior." *International Economic Review* 20 (February 1979): 29–58.

Vroman, Wayne

1973. "Serious Industrial Injuries and Worker Earnings in Florida." In *Supplemental Studies for the National Commission on State Workmen's Com-*

pensation Laws, vol. 2. Washington, D.C.: Government Printing Office.

World Health Organization

1980. *International Classification of Impairments, Disabilities and Handicaps.* Geneva: WHO.

Worrall, John D.

1982. "Overlapping Benefits: Workers' Compensation and Other Income Sources." Mimeo. New York: National Council on Compensation Insurance, November.

Worrall, John D., and Appel, David

1982. "The Wage Replacement Rate and Benefit Utilization in Workers' Compensation Insurance." *Journal of Risk and Insurance* 49, 3: 361–71.

Yelin, Edward; Nevitt, Michale; and Epstein, Wallace

1980. "Towards an Epidemiology of Work Disability." *Millbank Memorial Fund Quarterly* 58, 3: 386–415.

INDEX